WINE
A TASTING COURSE

WINE
A TASTING COURSE

FROM GRAPE
TO GLASS

MARNIE OLD

CONTENTS

NAVIGATING WINE BY STYLE

MASTERING WINE VARIABLES

DISCOVERING WINE GRAPES AND REGIONS

FOREWORD

FOUR GENERATIONS OF MY FAMILY in California have dedicated themselves to bringing the joys of wine to a wider audience. Growing up in this business, it has always been clear to me that people are very interested in wine, but they're also very much afraid of it. The biggest hurdle for wine lovers to overcome in learning about wine isn't necessarily the lack of information but, rather, a lack of trust in their own palates. People usually don't realize that their taste buds work just as well as those of a wine expert. If a wine tastes good and smells good to you, there's a 98 percent chance it is a good wine.

WHAT I FIND MOST COMPELLING about this book is that it addresses this issue head-on by helping people learn to trust their own palates right away. While simple in title, *Wine: A Tasting Course* is revolutionary and refreshingly direct in its design and approach. It focuses on sharing powerful ideas and practical skills, rather than excessive information, with creative, understandable graphics and straightforward descriptions to encourage wine drinkers to explore at their own pace. Lots of real-world examples and "try at home" interactive exercises are included to help people discover their own wine *aha!* moments. The beauty of this book is that you can choose to read it as an in-depth educational tome or to skim it for insights and visual references that decode challenging concepts, removing the mystery of wine without taking away the magic.

AS A WINEMAKER, I used to think I was adding character and personality to wine, but I soon learned that my work was more like being a babysitter than being a wine "maker." Mother Nature is the winemaker; we just need to stay out of her way by finding the right soil and climate, then manicuring the vines to help them reach their full flavor potential. I see a similar pattern in this book's uncommonly simple approach to teaching wine. By focusing on a handful of central truths about how wine really works, it breaks wine down without dumbing it down. *Wine: A Tasting Course* shows the reader the big picture and then stays out of the way, encouraging everyone who picks up the book to explore and enjoy wine on their own terms.

WHEN I FIRST MET MARNIE OLD, more than 20 years ago, she was a young sommelier, and I was impressed with her zeal, commitment, and desire both to learn and to teach about wine at every turn. It is exciting to see her grow into a leading voice in the wine world and for her vision to culminate in this terrific book. Whether you are a wine aficionado or a novice wanting to learn more, this is a must-read for any lover of the grape.

Michael Mondavi

INTRODUCTION

If you enjoy wine, but find it confusing, this is the book for you.

You don't need to memorize reams of data to feel confident shopping for wine or deciding what to drink in restaurants. All you need to learn are a few powerful ideas, a handful of concepts that help explain wine's dramatic variations of style. Once you know *why* wines taste the way they do and *how* to zero in on the styles that suit your tastes, you'll feel more in control when buying.

Wine can seem terribly complex and impenetrable to beginners, but it becomes much less frustrating when we step back and look at the big picture. Most wine books miss the forest for the trees, providing oodles of wine details but little in the way of practical wine skills. *Wine: A Tasting Course* is different. This visual guide explains how wine works, sharing useful concepts that can help anyone navigate the wine world without feeling insecure or ill informed.

Instead of zooming in on a thicket of wine labels, *Wine: A Tasting Course* combines colorful images and infographics to quickly convey the kinds of practical generalizations wine professionals use to make educated guesses about how any given wine will taste. Instead of presenting wine as a wholly foreign topic, *Wine: A Tasting Course* uses what you already know to help you solve the wine puzzle. If you can picture how peaches change in flavor as they ripen, then you can easily grasp why wines from cooler regions taste mild and acidic and those from warmer places taste bolder and more dessert-like.

This is not to say that the traditional wine book is obsolete. No agricultural product is as diverse and rarefied as wine, and no commercial product can rival its arcane system of nomenclature. There will always be a need to "look things up," in thick reference tomes. But *Wine: A Tasting Course* takes a refreshingly different path. Instead of focusing on what sets each individual wine apart—cataloging grapes and regions, vintages and vintners—it explains what all wines have in common and suggests sensible ways to sort them into groups based on how they taste, not where they were made.

Wine has an unfortunate reputation for snobbery and pretentiousness, but the top wine professionals are almost never snobs. We love "serious" wines, of course, but recognize the need for "un-serious" wines, too. The true wine experts are those who have moved past wine knowledge toward wine enlightenment. With the confidence to make educated guesses based on a few central truths, wine insiders can relax and savor what's in the glass regardless of its origins. This book aims to provide a similar degree of comfort and ease with wine for all drinkers, to share expert-level ideas that can take the stiffness out of wine and put the fun back in. By teaching you how to trust your own senses, *Wine: A Tasting Course* will set you on your own path to *v*inlightenment and *v*independence.

USING THIS BOOK

Wine: A Tasting Course is a different kind of wine book, sharing professional-level insights in refreshingly direct terms. It assumes no prior wine knowledge, favors everyday language over wine-trade jargon, and provides guided comparative tastings to reinforce key concepts. What you'll find here are simplified versions of the useful generalizations and practical skills used by sommeliers and winemakers the world over. Since this book's goal is to smooth the path to wine comprehension and to provide the reader with information of real-world relevance, some of wine's complexities must be downplayed to serve the larger objective. This is done not to mislead but because, in every subject, we must learn to walk before we can run.

Within this book, consider the chapters as individual lessons—and, if possible, taste along with the pages labeled "The Tasting." Take your time, and have fun with those sections.

- The Tastings are self-guided samplings of two to four wines, designed to be executed at home. If you're pulling more than one cork, why not make it a social gathering? Consider starting your own wine-tasting group with 4 to 10 wine-loving friends. If this isn't practical and you need to taste with only one or two people, don't fret about waste; for a helpful tip on preserving opened wines to drink later, see Freezing Wine, p.63.

- For maximum global relevance, the wine recommendations given are broad and rely on widely distributed styles, but some may be hard to locate in your area. Luckily, many wines share similar characteristics. Explain the situation to your wine retailer and ask them to recommend a reasonable substitute.

- If home tasting isn't feasible, consider your local wine bar. Many of these tastings might be feasible with wines served by the glass.

- Of the wines available in the categories specified here, 80 percent or more should fit their sensory descriptions well enough to get across the point of each lesson. However, not all wines taste exactly alike. There are always a few that defy even the sommelier's expectations, and that's okay. Simply treat any exceptions you encounter as delicious digressions and try again.

BUILDING WINE SKILLS

DO YOU COUNT YOURSELF among the wine-curious but find you are stumped in the wine aisle? If just trying to explain what you want to drink leaves you tongue-tied, memorizing grape names and wine regions won't do much good. It's a little bit like learning to drive: The first step is learning practical skills such as the rules of the road and how to steer, not theoretical knowledge like how the internal combustion engine works. What wine lovers need above all is to get comfortable with real-world wine activities. Discovering what the wine world has to offer should be delicious fun. A few lessons can make all the difference, starting with the most relevant topics: how to taste wine and describe it, how to shop for wine, and how to get the most enjoyment from each bottle. Once you feel more in control on everyday wine tasks, you will be inspired to take the driver's seat and explore the wonderland of wine on your own terms.

TALKING
AND
TASTING

THE INSIDER'S SENSORY TOUR Wine is an amazing drink—but it is a drink that leaves many at a loss for words. Taking control of our wine experiences becomes much easier when we know what to look for as we taste and how to communicate about wine's variations in style. There is no need to adopt pretentious descriptive prose or to get lost in subjective perceptions. Wine's most fundamental qualities can be identified by taking one sense at a time. A simple sensory checklist enables us to evaluate new wines and to describe accurately what we like and don't like.

WINE LINGO AND GEEK SPEAK

Enjoying wine is easy, but communicating about it is difficult. What people really want to know is what wines taste like and how they differ from one another. Unfortunately, our day-to-day vocabulary is weak in the realm of smells and tastes—the areas in which wine excels.

Wine labels and restaurant wine lists rarely address the taste of wine, more often conveying ingredients and wine regions. The first step in making sense of wine is to learn how to talk about the experience of wine tasting and how to interpret what others say. For beginners, a handful of terms that describe wine's characteristics is all that's needed. The key is to approach the challenge in an organized way, with a sensory checklist of qualities that are reasonably objective.

FLINTY MEATY TURGID SMOOTH
RUSTIC CHEWY JAMMY BUTTERY
VOLATILE LEATHERY MINERAL
AUSTERE VEGETAL GRASSY
TIRED CHALKY ROBUST BRAMBLY
BARNYARD HOLLOW VANILLA
RACY UNCTUOUS

▲ **WINE WORDS**
These words crop up often in wine tastings. Feel free to use them, but add your own terminology, too.

THE IMAGINATIVE LANGUAGE USED IN WINE REVIEWS CAN BE DISORIENTING AND SOMETIMES EVEN UNAPPETIZING.

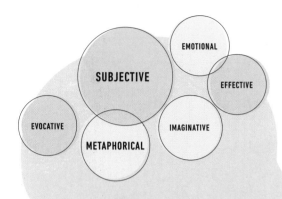

"THIS FRIENDLY, INKY SYRAH TASTES OF STEWED BOYSENBERRIES, WITH HINTS OF PENCIL LEAD AND FOREST FLOOR."

INDIRECT

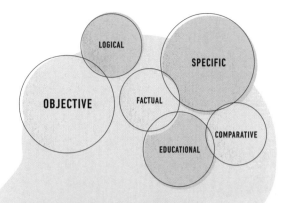

"THIS SAUTERNES IS FULL-BODIED AND SWEET, WITH STRONG OAK FLAVORS AND PLENTY OF BALANCING ACIDITY."

DIRECT

DEAL IN DESCRIPTORS

When we think of wine language, we tend to picture wine-label terms: grape names such as Chardonnay, or wine appellations like Bordeaux. But the most useful wine terms for the novice are descriptors. These words can help us describe the wine qualities we enjoy most and avoid those we don't. There are two main types of wine descriptors, which we will call indirect and direct. Neither is right or wrong. They are simply different ways of talking about wine.

INDIRECT WINE TERMS

Experts often paint a poetic "word picture" of a wine's flavor to convey complex ideas quickly to their audience. These terms:

- describe wine metaphorically by comparing it to other experiences
- describe subjective features that may be perceived differently by each individual
- are limitless in number and can include words that convey emotion and bias
- often attempt, through evocative language, to capture elusive olfactory scents and flavors
- are highly effective for motivating sales, as used in marketing and media
- are best suited to one-way communication from a wine professional

DIRECT WINE TERMS

To evaluate wines dispassionately, professionals use specific terminology for the most important of wine's sensory characteristics. These terms:

- describe wine's primary traits—like color, sweetness, and strength—typically on a power scale
- refer to objective features that most people perceive in the same way
- are limited in number, tending to be concrete terms that convey less personal bias
- pinpoint actual sensory qualities, such as how wine looks, tastes, smells, and feels
- are highly effective for comparative analysis, as used in winemaking and education
- help facilitate meaningful two-way communication

TASTE WINES LIKE A PRO

Our perception of wine is easily influenced by the environment in which we taste, so professionals try to be as objective as possible. A consistent tasting routine helps establish a baseline for comparison. The goal is to isolate and amplify the impact of wine's sensory characteristics—colors, scents, flavors—to distinguish one wine in a well-lit and odor-free room and following the steps below.

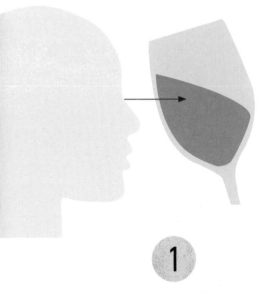

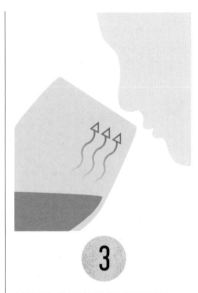

1

LOOK AT THE WINE

Is the wine white, rosé (pink), or red? If the tasting environment is well lit, tilt the glass over a white surface, such as a sheet of paper, and look through the wine for useful style clues. How deep is the color? Is it showing signs of browning with age?

See pp.20–23 for more on what to look for.

2

SWIRL THE WINE IN THE GLASS

We swirl wine to smell it better—almost like turning up the volume on the stereo system. A wine's smells grow more intense when its aroma compounds are concentrated in the bowl of a wine glass. Swirling increases the wine's surface area, which in turn boosts its rate of evaporation and aromatic intensity.

3

SNIFF THE WINE DEEPLY

Smell is the main sense used in wine tasting, so sniffing wine before tasting is an essential step. Dip your nose into the wine glass, and take two or three deep sniffs. Think about what you are smelling. How intense is the aroma? Does it remind you of anything? Fruits or vegetables? Herbs or spices? Do you smell toasty oak barrels?

See pp.24–25 for more on what to look for.

THE GOAL IS TO ISOLATE AND AMPLIFY THE IMPACT OF WINE'S SENSORY CHARACTERISTICS.

TO SPIT, OR NOT TO SPIT? THAT IS THE QUESTION

Wine professionals often spit wine at large tastings, something that seems unnatural to most people, since spitting is rude in any other context. However, for those who must taste wine critically as part of their work, spitting is essential, since it minimizes alcohol absorption and prevents intoxication. At large wine-tasting events, in winery tasting rooms, and in wine classes, spittoons are always readily available.

MODERN WINE-TASTING SPITTOON

4

SIP THE WINE

Take a slightly larger sip than usual. Instead of swallowing immediately, hold the wine in your mouth for 3–5 seconds, letting it coat every surface: tongue, cheeks, palate.

5

SWISH IT AROUND

By swishing wine around as if it were mouthwash, you can boost the sensory perceptions of taste, smell, and "mouthfeel" quite dramatically. Increasing surface contact makes tastes and tactile sensations more vivid. It also warms the wine; body heat increases wine's evaporation rate, intensifying its aromas for the olfactory nerves.

6

SAVOR THE WINE

Wine's flavor does not disappear when you swallow. Its aftertaste lingers for a minute or more, allowing you to reflect on and assess its sensory traits by ticking through the wine-tasting checklist. This is also the stage at which wine's quality can be evaluated. Decide whether you like the wine. Would you prefer it alone or with food? Would you buy it again?

See p.20 for the wine-tasting checklist.

WINE-TASTING CHECKLIST

Think of each new wine you taste as an entry in a mental database. Deciding how it compares to others you've tried determines where it gets classified. The final step of tasting is to savor, assessing the main qualities that are reasonably objective as we commit the wine to memory. At this stage, we'll use a checklist of sensory characteristics so we don't miss anything important.

USE YOUR SENSES— WELL, MOST OF THEM!

Four of our senses help us evaluate different wine attributes; the one sense that doesn't have any part to play in wine tasting is hearing. In the chart below, track any attribute across to the low, medium, or high column to find useful terms for describing how it manifests itself in a wine's style.

PUT IT IN THE VAULT
Describing a wine in words, even just to yourself, is the key to remembering its characteristics so you can compare them to those of wines you'll taste in the future.

SENSE	ATTRIBUTE	LOW	MEDIUM	HIGH
See	COLOR	White	Pink	Red
	COLOR DEPTH	Pale	Moderate	Dark
Taste	SWEETNESS	Dry	Lightly sweet	Fully sweet
	ACIDITY	Low acid	Tangy	Tart
Smell	FRUIT INTENSITY	Mild	Flavorful	Bold
	OAK PRESENCE	No oak	Mild oak	Strong oak
Feel (mouthfeel)	BODY	Light	Mid-weight	Heavy
	TANNIN (RED ONLY)	Silky	Velvety	Leathery
	CARBONATION	Still	Spritzy	Sparkling

HOW WINE LOOKS

The most obvious differences between wines are the ones we can see. People are so naturally focused on visual perception that wines on wine lists and in retail stores are usually classified by color.

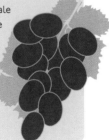

FIGURE OUT THE COLORS

Wine colors range from nearly clear to inky purple-black, but the first step is to decide into which broad category they fit: white, pink, or red. A few rare wines are made in a specialized way that obscures their original color; these are mostly gold- or amber-toned sweet wines that are made from dried grapes or brassy "orange wines" that are deliberately oxidized during winemaking. However, 99.9 percent of the time, it will be quite obvious which wines are white, which are red, and which fall somewhere in between.

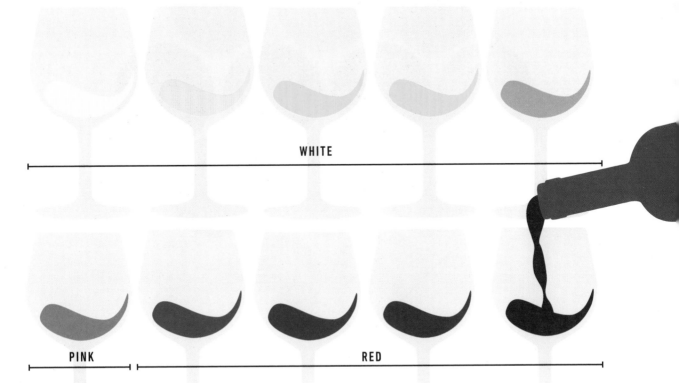

WHITE

PINK

RED

RARE ROSÉS
Only a tiny percentage of wines fall into the pink middle ground of color. These are called rosé wines, using the French word for "pinked."

ASSESSING WINE BY COLOR DEPTH

The intensity of color within each of the white, red, and rosé categories can give the drinker a hint about how a wine will taste. As a general rule, color saturation tends to match flavor intensity, and it can also provide clues on other qualities. White wines with a golden hue, for example, are more likely to be oaky than those that are almost water white; and pale, translucent red wines are often lighter-bodied and less astringent than darker reds.

FRESH OR FADED?
Red wines grow paler over time. Their colors shift from pinky purple in youth, to a browner rusty orange in advanced age.

SHOWING SOME SKIN
Red wines are darker when they are made from grapes with thicker skins and smaller berries or when they have had longer grape-skin contact during fermentation. Grape skins supply not only color but also flavor, so reds and rosés tend to taste bolder than whites.

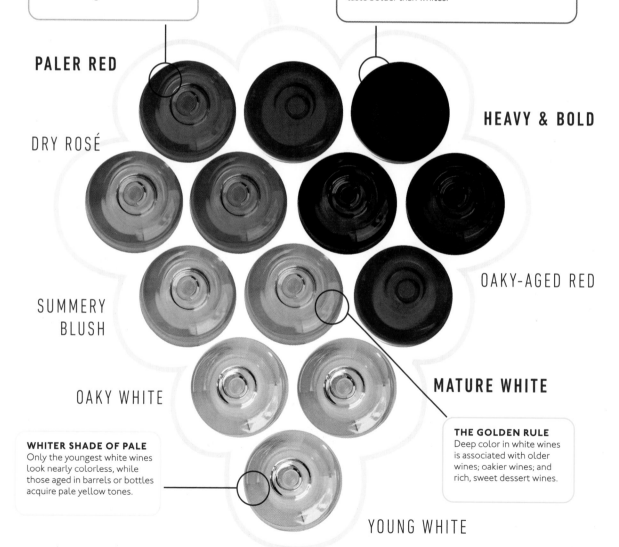

PALER RED

HEAVY & BOLD

DRY ROSÉ

OAKY-AGED RED

SUMMERY BLUSH

MATURE WHITE

OAKY WHITE

THE GOLDEN RULE
Deep color in white wines is associated with older wines; oakier wines; and rich, sweet dessert wines.

WHITER SHADE OF PALE
Only the youngest white wines look nearly colorless, while those aged in barrels or bottles acquire pale yellow tones.

YOUNG WHITE

WHITES: WHAT AFFECTS THEIR COLOR?

The main color source in white wines is oxidation: Exposure to air deepens whites from faintest yellow to gold. The most common source of oxidation is barrel aging, so oaky Chardonnay tends to be darker and more golden than crisp, stainless-steel–fermented Sauvignon Blanc. Uncommonly intensely flavored white wines will also display more color saturation, as is the case with sweet dessert wines.

DARKENED BY BARRELS, AGE, OR DENSITY

REDS: WHAT AFFECTS THEIR COLOR?

Just like whites, red wines look darker when they are more concentrated. However, color behaves quite differently in red wine because of its source: dark grape skins. While white wines darken with age, reds do the reverse, growing paler as their color compounds succumb to gravity, settling as sediment.

The type of grape, degree of ripeness, and techniques used to extract color from the skins all affect a red wine's color depth. Thin-skinned grapes like Pinot Noir make paler wines than those from thicker-skinned varieties like Syrah; and fruit from sunny regions provides deeper color than cooler-climate fruit.

Vintners extract more color from the grape skins for premium age-worthy reds, while the hue of rosés is determined by limiting contact with the grape skins.

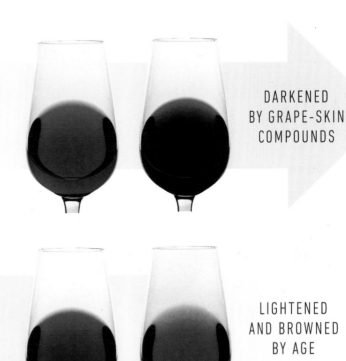

DARKENED BY GRAPE-SKIN COMPOUNDS

LIGHTENED AND BROWNED BY AGE

ARE WE TASTING OR SMELLING?

In everyday speech, we use the word "taste" for all sensations happening in the mouth. In the wine world, too, "taste" is mostly used in this generalized context. However, when we analyze wine, we make distinctions between wine characteristics based on which sense perceives them.

TAKE A BIG SNIFF

Wine experts spend as much time sniffing wine as drinking it. This is because a wine's scent provides an almost perfect preview of its flavor, and one we can enjoy without employing our glass.

KNOW WHAT TASTE IS

By separating the three sensory threads—smell, taste, mouthfeel—that occur almost simultaneously when we take a sip, we can distinguish true tastes from the smells that constitute "flavor" and from the tactile sensations known as mouthfeel. For example, we might say that crème brûlée tastes sweet and creamy, or like vanilla and caramel. But in sensory science, only sweetness would be considered a true taste because it alone is detected by the tongue's taste buds. Vanilla and caramel "flavors" are really olfactory sensations, or smells, while creaminess is a tactile sensation, part of the dessert's mouthfeel.

PERCEPTION OF WINE
As you drink wine, its true tastes register on the tongue and its mouthfeel is felt on the fleshy parts of your mouth. However, wine's smells are perceived through both the nose and the mouth.

EXTERNAL SMELLS REGISTER AS **ODORS**

SMELLS
Olfactory nerves detect odors and flavors, such as wine's fruit and oak components.

INTERNAL SMELLS REGISTER AS **FLAVORS**

MOUTHFEEL
The flesh of the tongue, palate, cheeks, and gums detects tactile sensations, such as texture, carbonation, and tannic astringency.

TASTES
Taste buds on the tongue can detect only limited taste sensations, such as sweetness and acidity.

UNDERSTAND AROMAS

Of all our senses, smell is the most important when it comes to wine tasting. Even if we don't sniff our wine, we still get an intense blast of aromas when we take a sip. We're used to thinking of this as flavor, as part of how wine tastes, but most of what we perceive as flavor is really olfactory stimulus—or smells.

Technically, there is no real difference between odors and flavors, except the direction from which they arrive. Olfactory nerves in the upper nasal cavity recognize smells as odors when sniffed through the nose from an external source. However, when those same smells reach the nose from the internal passage that connects the nose and mouth, they register as flavors, as part of how food or drink tastes.

Wine descriptors start to make a lot more sense once we grasp the fundamental difference between the taste sensations that are conveyed by the taste buds and the smell sensations delivered by the olfactory nerves. We taste a few rudimentary wine qualities—such as sweetness and acidity—on contact with our tongue. But we perceive a great many more complex wine characteristics as both scents and flavors when the volatile aroma compounds in wine reach our olfactory nerves.

HUMANS CAN DISCERN AT LEAST 10,000 DIFFERENT SMELLS ...

PASS THE SMELL TEST

For a vivid illustration of the difference between taste and smell, try this little experiment.

- Plug your nose tightly, and take a sip of orange juice. Keep your nose blocked all the while, and don't let go for at least 5 seconds after you swallow.
- Notice how, while the nasal passage is blocked, you can discern only what your tongue can taste on its own—in this case, sweetness and acidity.
- Now unplug your nose. As soon as air can flow freely from your palate to your olfactory nerves, you will get a rush of citrusy orange "flavor."

... BUT ONLY SIX SENSATIONS CAN BE DETECTED WITH THE TONGUE ALONE.

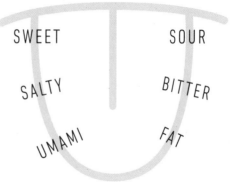

SWEET   SOUR

SALTY   BITTER

UMAMI   FAT

HOW WINE TASTES

Now that we know there are only six detectable taste sensations, we need to establish what impact that has on wine tasting. The most surprising thing is probably that only two "true" tastes are of significance.

ONLY TWO TASTE SENSATIONS ARE IMPORTANT IN WINE TASTING: SWEETNESS AND ACIDITY.

KNOW THE SENSATIONS

Four of the six known taste sensations have been recognized for centuries: sweetness, sourness (or acidity), saltiness, and bitterness. The other two are far less apparent and were only discovered with more recent laboratory testing. The overall "yummy," or savory, quality called umami, triggered by glutamates and amino acids, was first identified by Japanese researchers curious as to why seaweed and miso tasted so satisfying. More recently, another barely detectable taste has been found to be associated with fat in foods.

WHAT TO LOOK FOR

Sweetness and acidity are the two tastes we look for and assess whenever we taste wine, and both are important in categorizing wine by style. We do not look for other tastes because wine has no salt, fat, or bitterness—not tongue bitterness, like that found in beer or coffee. While many wines feature umami, it's not readily apparent.

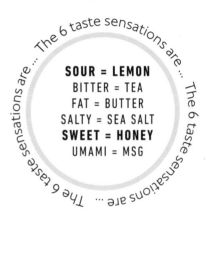

... The 6 taste sensations are ... The 6 taste sensations are ... The 6 taste sensations are ... The 6 taste sensations are ...

SOUR = LEMON
BITTER = TEA
FAT = BUTTER
SALTY = SEA SALT
SWEET = HONEY
UMAMI = MSG

GAUGING SWEETNESS

In the wine world, sweetness is measured in grams of sugar per liter of wine (g/l). This diagram gives an idea of how wines compare with some other beverages.

DRY MERLOT
2g/l

LIGHTLY SWEET
RIESLING 15g/l

TEA WITH ONE
SUGAR 12g/l

SWEET OR DRY?

Sweetness is perceived as a sugary sensation on contact with the tongue, most vividly at the very tip, where taste receptors are more densely concentrated. Most wines have no perceptible sweetness and are described as "dry." This tends to be a confusing descriptor for beginners, because dry has a different meaning in wine than it does in everyday use. For centuries, winemakers around the world have called wines dry when their natural grape sugar has been fully converted to alcohol. *Sec* in French, *trocken* in German, and *secco* in Italian all mean "not wet" in normal speech but "not sweet" when applied to wine.

A pleasing hint of sweetness can be found in varying degrees in wine, most often in mass-market bargain wines. Lightly sweet "off-dry" styles are particularly popular with wine novices, who appreciate their juice-like flavor. Fully sweet wines, or dessert wines, are seductive but rare because they are challenging and expensive to produce. The vast majority of the world's wines are dry because they are simpler to make, have a longer shelf life, and work well with food.

HIGHS AND LOWS

This chart shows common terms for low-, medium-, and high-sugar wines, along with details of how each is perceived on tasting and examples of corresponding wines.

SUGAR	TERM	DESCRIPTION	WINE EXAMPLE
Low	Dry	No noticeable presence of sugar	Australian Chardonnay; French Côtes du Rhône
Medium	Lightly sweet; off-dry	Noticeable presence of sugar	German Riesling; California Zinfandel–based blends
High	Sweet; dessert	Obvious, strong presence of sugar	Portuguese Port; Italian Moscato

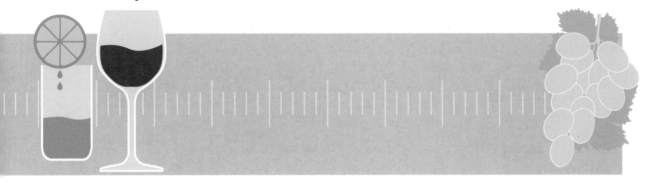

SWEET PORT
100g/l

ORANGE JUICE
85g/l

WINE-GRAPE
JUICE 225g/l

SOUR FRESH-FRUIT ACIDITY

Acidity is perceived as a sour sensation on contact, causing the mouth to salivate almost immediately, as with lemon juice or vinegar. Wine is more tart than most beverages, due to the high acidity of fresh grapes.

Newcomers often find wines too sour for their liking, in part because acidity always seems strongest on the first sip. But wine's acidic edge melts away as you continue to drink, especially if you're also eating. Since high levels of acidity can be a turnoff for the inexperienced, the wine profession treads carefully in describing it. Words such as "sour" and "acidic" carry negative connotations, so terms that sound more appetizing are more commonly used: tart or tangy, crisp or quenching, racy or refreshing.

A CHEF'S SECRET

Acidity makes everything it touches taste better. Chefs know this, and that's why so many restaurant dishes are made with a splash of wine, a squeeze of lemon, or a drizzle of vinegar.

HIGHS AND LOWS

This chart shows common terms for low-, medium-, and high-acidity wines, along with details of how each is perceived on tasting and examples of corresponding wines.

ACIDITY	TERM	DESCRIPTION	WINE EXAMPLE
Low	Mildly acidic; flabby	Noticeable but modest acidity, as in baked apples	Oaky Chardonnay; Cream Sherry
Medium	Tangy; crisp	Standard refreshing acidity, as in fresh apples	Italian Pinot Grigio; Chilean Merlot
High	Tart; sharp	Prominent, aggressive acidity, as in underripe apples	French Sancerre; Italian Chianti

ACIDITY PH SCALE

The pH scale measures acidity, albeit in a different way than you might expect. The lower the number, the more acidic the sample, with water—a neutral point between acid and alkali—having a pH of 7.

WINE
pH 3–4

VODKA
pH 6–7

LIME JUICE
pH 2

ORANGE JUICE
pH 3.5

COFFEE
pH 5.5

WATER
pH 7

ANTACID
pH 10

THE TASTING

Identifying Sweetness and Acidity

COMPARE FOUR WINES AT HOME

Sample the four white wine styles shown below.
1 Pay special attention to sensations on initial contact with your tongue.
2 Evaluate their levels of sweetness and acidity on a scale from low to high.
3 Consider which wine you prefer, whether alone or with food.

> **WASTE NOT WANT NOT**
>
> If you need to taste with only one or two people, don't fret about waste; for a helpful tip on preserving opened wines to drink later, see Freezing Wine, p.63.

1

LOW SWEETNESS, HIGH ACIDITY

2

LOW SWEETNESS, MEDIUM ACIDITY

3

MEDIUM SWEETNESS, HIGH ACIDITY

4

HIGH SWEETNESS, LOW ACIDITY

FRENCH SAUVIGNON BLANC

For example …
Sancerre, Pouilly-Fumé, Bordeaux Blanc, or Touraine Sauvignon Blanc

Can you taste …?
Very dry: very obvious absence of sugar

Tart acidity: uncommonly high degree of sourness

CALIFORNIA CHARDONNAY

For example …
Barrel-fermented styles from Sonoma, Monterey, or Central Coast

Can you taste …?
Dry: absence of sugar

Crisp acidity: standard degree of sourness

WASHINGTON RIESLING

For example …
Low-alcohol styles from Columbia Valley (particularly in German-style bottles)

Can you taste …?
Lightly sweet or off-dry: noticeable presence of sugar

Tart acidity: high degree of sourness

FRENCH MUSCAT VIN DOUX NATUREL

For example … Dessert wines, such as Muscat de Minervois or Muscat de Beaumes-de-Venise

Can you taste …?
Sweet: obvious presence of sugar

Low acid: uncommonly low degree of sourness

HOW WINE SMELLS

In wine tasting, "fruit" is the collective term for olfactory scents and flavors that come from the grapes used to make wine. Since all wines are made with 100 percent fruit, almost all wine flavor fits this description.

KNOW YOUR FRUIT FROM YOUR OAK

It is helpful to categorize wine's scents and flavors and assess their overall intensity. In wine tasting,

the two main categories of wine's olfactory sensations are "fruit" smells and "oak" smells, each named for their source in the winemaking process.

OF ALL WINE'S SMELLS, MOST COME FROM THE GRAPES.

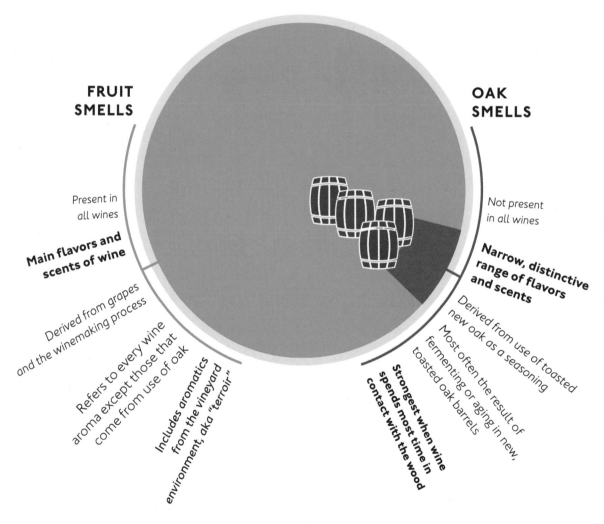

FRUIT SMELLS

Present in all wines

Main flavors and scents of wine

Derived from grapes and the winemaking process

Refers to every wine aroma except those that come from use of oak

Includes aromatics from the vineyard environment, aka "terroir"

OAK SMELLS

Not present in all wines

Narrow, distinctive range of flavors and scents

Derived from use of toasted new oak as a seasoning

Most often the result of fermenting or aging in new, toasted oak barrels

Strongest when wine spends most time in contact with the wood

HIGHS AND LOWS

This chart shows common terms for wines with low-, medium-, and high-intensity fruit, along with details of how each is perceived on tasting and examples of corresponding wines.

SCENT	TERM	DESCRIPTION	WINE EXAMPLE
Low intensity	Mild; subtle	Understated flavor, as with chamomile tea	Italian Prosecco; French Chablis—rare outside of white wines
Medium intensity	Moderate; flavorful	Standard flavor power, as with black coffee or tea	New Zealand Sauvignon Blanc; Spanish Garnacha
High intensity	Bold; concentrated	Intense, powerful flavor, as with espresso	Napa Valley Cabernet Sauvignon, German Eiswein—rare outside of red wines and dessert wines

UNDERSTANDING "FRUIT"

Many wine tasters try to identify specific aspects of a wine's fruit aromatics, with comparative terms like blackberry or lemon. For beginners, though, it is more useful to start by assessing the overall intensity of a wine's fruit on a simple power scale. Think of the fruit component as wine's flavor intensity: Is it subtle and understated, or bold and over the top?

Professionals in the wine trade stretch the word "fruit" beyond its normal meaning. Fruit can encompass plenty of overtly fruit-like smells, such as the pineapple scent of Chardonnay or the black currant aroma of Bordeaux. But in wine lingo, fruit is also a catchall category that encompasses non-fruit flavors, too. Whether it's the peppery scent of Syrah, the herbal flavor of Sauvignon Blanc, or a floral aroma in Moscato, these would all be described as part of the wine's fruit component. However, when a wine is described as "fruity," it typically means that its scent is strong and dominated by ripe dessert-like smells of actual fruits.

In addition to the natural smells present in the raw grapes, the fruit component incorporates smells that are generated during winemaking. Complex chemical reactions create volatile aromatic compounds that can smell like all sorts of things—from bread dough, to leather; from cedar, to asphalt.

EARTH BECOMES "FRUIT"

Vineyard environments can contribute earthy aromatics, often called *terroir*, that are considered dimensions of a wine's fruit component.

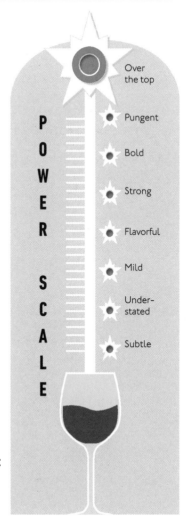

Over the top

Pungent

Bold

Strong

Flavorful

Mild

Under-stated

Subtle

FRUIT-INTENSITY POWER SCALE
Wines range widely in their flavor intensity. As you taste each new wine, try to place where it would score in terms of flavor per square in (sq cm).

WHY OAK?

For centuries, oak barrels and vats were essential for fermenting and storing wine, since oak is a dense wood capable of containing liquids. They continue to be used today for the desirable vanilla-like flavors they impart.

UNDERSTANDING "OAK"

The term "oak" describes a narrow range of wine scents and flavors, derived from contact with wood, usually toasted oak barrels. However, many wines are made without oak treatment, so these scents and flavors are not always present.

Oakiness is most commonly found in red wines and full-bodied whites, and it is particularly associated with premium aged styles. Barrels impart the most oak flavor when they are newest, so wines aged in older barrels may not show a trace of oakiness.

NEW OAK: SMELLS LIKE BROWN SPIRITS

New toasted oak gives wine a range of flavors that closely resemble those found in oak-aged spirits like cognac and bourbon. New oak is naturally high in aromatic compounds that resemble those in vanilla and dessert spices. The toasting process that is used to bend the wood into rounded shapes produces nutty lactones and caramelizes its surface. Winemakers use oak as a seasoning for wine and pay close attention to the type of oak used and how it is toasted.

For vintners, French and American oak are as distinct as Ethiopian and Colombian coffee. The degree of "toast" that the wood is given has a similar effect to the type of roast those coffee beans might receive. Smaller barrels also guarantee a stronger oak flavor in wine for the same reason that a fine espresso grind increases coffee's intensity: More surface contact between wine and oak means more flavor is transferred.

COGNAC
Intense toasted French oak flavor

BOURBON
Intense charred American oak flavor

OAKED WINE
Subtle toasted oak flavor

HIGHS AND LOWS

This chart shows common terms for wines with low, medium, and high levels of oak, along with details of how each is perceived on tasting and examples of corresponding wines.

OAKINESS	TERM	DESCRIPTION	WINE EXAMPLE
Low	Unoaked; naked	Absence of oak scents and flavors, as with vodka	German Riesling; Italian Valpolicella
Medium	Mild oak	Mild oak scents and flavors, as with young Irish whiskey	French Bordeaux; Oregon Pinot Noir
High	Oaky; toasty	Intense oak scents and flavors, as with premium aged cognac	Prestige Chardonnay; Spanish Rioja

THE TASTING

Identifying Fruit and Oak

COMPARE FOUR WINES AT HOME

Sample the four wine styles shown below.
1 Pay special attention to the scents and flavors present before and after taking a sip.
2 Try to distinguish the overall wine smell, or "fruit," from the toasty scent of oak.
3 Evaluate the intensity of each on a scale from low to high.

WASTE NOT WANT NOT

If you need to taste with only one or two people, don't fret about waste; for a helpful tip on preserving opened wines to drink later, see Freezing Wine, p.63.

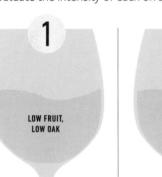

1 LOW FRUIT, LOW OAK

2 MEDIUM FRUIT, MEDIUM OAK

3 MEDIUM FRUIT, HIGH OAK

4 HIGH FRUIT, HIGH OAK

UNOAKED FRENCH CHARDONNAY

For example ...
Mâcon-Villages, Chablis, St-Véran, Bourgogne Blanc, or Viré-Clessé

Can you taste ...?
Subtle fruit: low-intensity wine flavor

No oak: no detectable scent of oak barrels

BARREL-FERMENTED CALIFORNIA CHARDONNAY

For example ...
Styles from Sonoma, Central Coast, Monterey, Santa Barbara, or Carneros

Can you taste ...?
Moderate fruit: medium-intensity wine flavor

Mild oak: noticeable scent of oak barrels

BARREL-AGED SPANISH TEMPRANILLO

For example ...
Rioja, Toro, or Ribera del Duero of the *crianza* or *reserva* classifications

Can you taste ...?
Moderate fruit: medium-intensity wine flavor

Strong oak: overt scent of oak barrels

BARREL-AGED AUSTRALIAN SHIRAZ

For example ...
Premium styles from Barossa, McLaren Vale, or South Australia

Can you taste ...?
Bold fruit: high-intensity wine flavor

Strong oak: overt scent of oak barrels

HOW WINE FEELS

Some sensory aspects of how wine "tastes" are neither tastes nor smells but "mouthfeel"—tactile sensations we discern with tongue, palate, lips, and gums. Mouthfeel encompasses some of our favorite food pleasures—from the crispy crunch of potato chips to the creaminess of chocolate mousse.

IN WINE, WE LOOK FOR THREE TYPES OF PHYSICAL SENSATIONS:

1. CARBONATION, OR BUBBLES
2. WEIGHT, OR BODY
3. TANNIN, OR ASTRINGENCY

FIZZ AND FROTH

The most obvious and instantaneous tactile sensation in wine is carbonation, since bubbles are immediately apparent on contact. Carbon dioxide is a natural by-product of fermentation, so all wines are bubbly at some point during winemaking. This natural fizz is usually allowed to escape, leaving the vast majority of wines "still," or non-carbonated. Sometimes, though, the bubbles are intentionally trapped and preserved to make a wine that is fully carbonated, or "sparkling."

Sparkling wines foam in the mouth like a soda and deliver a similar tingly shiver of refreshment on the palate. They need special bottles and corks to contain their carbonated contents under pressure. Occasionally, a milder prickle of carbonation, known as a spritz, may be present in wines that are not fully sparkling; these wines tend to be very young whites or rosés. Their bubbles typically dissipate quickly in the glass.

HIGHS AND LOWS

This chart shows common terms for wines with low, medium, and high levels of carbonation, along with details of how each is perceived on tasting and examples of corresponding wines.

TEXTURE	TERM	DESCRIPTION	WINE EXAMPLE
Low	Still; standard	Total absence of bubbles or fizz	Pinot Gris; Sauvignon Blanc
Medium	Spritzy	Faint presence of bubbles or fizz	Vinho Verde; Basque Txakolina
High	Sparkling; bubbly	Overt presence of bubbles or fizz	Champagne; Prosecco

ASSESSING WEIGHT

Weight in wine terms refers to texture or thickness—a physical sensation of viscosity on the palate that is also known as "body." In the same way that cream feels thicker than milk because it contains more fat, full-bodied wines feel heavier than light-bodied wines because they contain more alcohol. Most mid-weight wines have roughly 13.5% alcohol. The lower a wine's strength, the lighter it feels on the palate.

Dessert wines are an exception to this rule, since dissolved sugar adds viscosity of its own. Wines that are both sweet and strong, like Port, are very heavy, feeling nearly as thick as syrup. Other factors, such as oak aging or yeast contact, can boost perceived weight but usually to a lesser extent than alcohol and sugar.

FULLER-BODIED

▼ WHITER, LIGHTER
Most white wines tend to be lower in alcohol and feel lighter than most red wines, but there are a few exceptions to the rule.

13.5%

▲ SCARCE REDS
Red wines are rarely made with less than 12.5% alcohol, so truly light reds are hard to find.

LIGHTER-BODIED

CHECK OUT THOSE LEGS

A wine's weight becomes visible when wine slides down the sides of a glass after swirling. Known as "legs" or "tears," these drips fall slowest in the heaviest wines.

HIGHS AND LOWS

This chart shows common terms for low-, medium-, and high-texture wines, along with details of how each is perceived on tasting and examples of corresponding wines.

TEXTURE	TERM	DESCRIPTION	WINE EXAMPLE
Low	Light; light-bodied	Sheer, delicate texture, as with skim milk	German Riesling, Italian Moscato—rare outside of sparkling, white, and rosé wines
Medium	Mid-weight; medium-bodied	Standard medium texture, as with chocolate milk	French Bordeaux Blanc, Chilean Merlot
High	Heavy; full-bodied	Rich, viscous texture, as with chocolate milkshake	California Old-Vine Zinfandel, Portuguese Port—rare outside red, dessert, and fortified wines

DISCOVERING TANNIN

Red wines often dry out the mouth in the minute after tasting, blocking salivation and leaving a rough, leathery feeling on the palate. This is due to the presence of tannin, an astringent phenolic compound found in the skins, seeds, and stems of grapes as well as in new oak barrels.

- We find significant tannin only in red wines, because reds are fermented in contact with these solids, whereas white wines are not.
- Tannins add to a red wine's depth of color and also tend to coincide with flavor intensity.

- Tannin is a strong antioxidant and natural preservative that helps wines age but breaks down over time.
- The youngest, darkest, most intense red wines designed for aging are typically the most tannic.
- Sometimes called a wine's "grip," tannin's mouth-drying effect is not always immediately apparent but gets stronger in the 30–60 seconds after tasting.
- Wines with mild tannin can feel plush, like velvet, while strong tannin can leave a harsher feeling, like suede.

MOUTH-DRYING, NOT DRY

Tannin is often confused with dryness because it makes the tongue feel dry. However, in wine lingo, "dry" refers to wines that are not sweet, while wines that dry out the mouth are described as tannic.

RED WINES GET THEIR TANNIN MAINLY FROM GRAPE SKINS. THESE ANTIOXIDANT COMPOUNDS GIVE RED WINE COLOR AND FLAVOR, AS WELL AS ASTRINGENCY.

HIGHS AND LOWS

This chart shows common terms for low-, medium-, and high-tannin wines, along with details of how each is perceived on tasting and examples of corresponding wines.

TANNIN	TERM	DESCRIPTION	WINE EXAMPLE
Low	None	No detectable mouth-drying astringency	French Beaujolais; Dry rosé
Medium	Velvety; soft tannin	Detectable, gentle mouth-drying astringency	California Merlot; French red Burgundy
High	Leathery; hard tannin	Obvious, aggressive mouth-drying astringency	Italian Barolo; Australian Cabernet Sauvignon

THE TASTING

Identifying Body, Tannin, and Carbonation

COMPARE FOUR WINES AT HOME

Sample the four wine styles shown below.
1 Pay special attention to how they feel in the mouth.
2 Look for weight, or thickness of texture, and for the telltale prickle of carbonation.
3 After swallowing the reds, consider the astringent, mouth-drying effect of tannin.

WASTE NOT WANT NOT

If you need to taste with only one or two people, don't fret about waste; for a helpful tip on preserving opened wines to drink later, see Freezing Wine, p.63.

1

LIGHTWEIGHT,
NO TANNIN,
HIGH CARBONATION

2

LIGHTWEIGHT,
NO TANNIN,
MEDIUM CARBONATION

3

MID-WEIGHT,
MEDIUM TANNIN,
NO CARBONATION

4

HEAVYWEIGHT,
HIGH TANNIN,
NO CARBONATION

ITALIAN PROSECCO

For example …
Simple affordable brands

Can you detect …?
Light: delicate, sheer texture

Sparkling: vigorously carbonated

PORTUGUESE VINHO VERDE

For example …
Simple affordable brands

Can you detect …?
Light: delicate, sheer texture

Spritzy: faintly carbonated

NEW ZEALAND PINOT NOIR

For example …
Moderately priced brands

Can you detect …?
Mid-weight: standard medium texture

Soft tannin: mildly mouth-drying

Still: not carbonated

CHILEAN CABERNET SAUVIGNON

For example …
Premium or moderately priced brands

Can you detect …?
Heavy: dense, rich texture

Strong tannin: overtly mouth-drying

Still: not carbonated

EVALUATING WINE QUALITY

One of the most difficult things for wine novices is figuring out how to assess the technical quality of a wine, rather than simply determine whether or not it suits their personal tastes.

ANALYZE THE FINISH

Of all quality indicators, the length of a wine's "finish" is the most important and the easiest to discern. The term refers to the period of time after swallowing a wine during which its flavor sensations continue to resonate in the mouth; it's a fancy word for aftertaste. Lingering sensory impressions can last for anywhere from 30 seconds to 5 minutes, and their duration is a reflection of the wine's quality.

Wines made with impeccable ingredients and craftsmanship consistently display a longer, stronger, more pleasant finish than is found in less ambitious wines. A wine's finish is also where the damaging effects of poor storage, faulty corks, or winemaking mistakes become apparent.

A fine wine's finish is perceived not simply as lasting tastes, smells, and mouthfeel but as an almost palpable vibrating energy that hums in the mouth after tasting. The finish of serviceable bargain wines will fade fastest; well-made wines with higher quality standards will hang around in the mouth for at least 1–2 minutes; and truly exceptional wines can dominate the palate for considerably longer. During this period, experts pay close attention, because each wine's finish will ebb and flow on the palate in a unique way.

THE QUALITY OF A WINE CAN BE MEASURED BY THE LENGTH OF TIME IT RESONATES IN THE MOUTH.

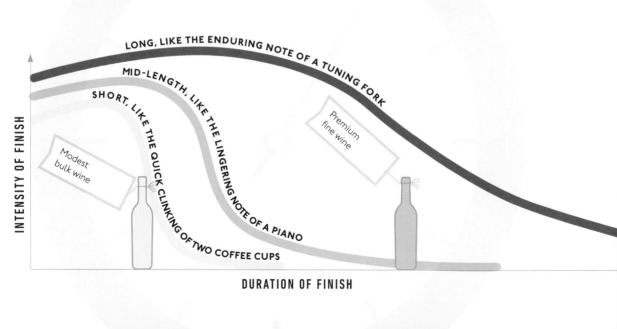

LONG, LIKE THE ENDURING NOTE OF A TUNING FORK

MID-LENGTH, LIKE THE LINGERING NOTE OF A PIANO

SHORT, LIKE THE QUICK CLINKING OF TWO COFFEE CUPS

Modest bulk wine

Premium fine wine

INTENSITY OF FINISH

DURATION OF FINISH

ASSESS CORK TAINT

Wine bottles have been sealed with cork for centuries, but many vintners are now going cork-free because natural corks can sometimes degrade wine's flavor. Around 5 percent of cork-sealed wines are noticeably contaminated by their cork, but the degree of damage varies widely. So-called "corked" wine has a characteristic fungus scent. In the worst cases, it can smell distinctly unpleasant, with odors reminiscent of moldy cardboard or mildewed wool. In mild cases, though, it can be more apparent from the absence of good wine smells than the presence of bad ones. This insidious problem is hard to recognize unless multiple bottles of the same wine are compared side by side. The wine trade is well aware of this issue, so when in doubt, speak up. Restaurateurs and retailers should replace suspect bottles in most instances.

IDENTIFY HEAT DAMAGE

As with fruit, wine keeps longer when refrigerated. Wines all go bad eventually, but the colder they are, the more slowly they change. Heat increases the rate of the chemical reactions that take place as wine matures, most notably oxidation. Wine ages prematurely when kept too warm, leaving it seeming cooked—browned in color and flattened in flavor. Extreme heat can alter the internal pressure of wine bottles, compromising their seal, which further speeds oxidation. Evidence of this is sometimes visible in seeping bottles or raised corks, but heat damage is usually less obvious. Short bursts of heat or longer storage at room temperature can permanently shorten a wine's potential life without showing obvious signs until it is tasted, showing in a short, dull, and lifeless finish. The optimum storage temperature for wine, to reduce spoilage as far as possible, is in the range of 50–60°F (10–15°C).

Exceptional luxury wine

MEASURING QUALITY
The old adage that you get what you pay for holds perfectly true for the intensity and length of finish in a luxury wine. This is one of the key elements lacking in bargain wines.

CHAPTER CHECKLIST
Here is a recap of some of the most important points you've learned in this chapter.

- Wine **descriptors** help us explain wine qualities. They can be **indirect** and evocative, or **direct** and dispassionate.
- We evaluate wine attributes through the use of all of our **senses** except hearing.
- Most wine **color** comes from **grape skins**. White wines pick up no color, but the color of rosés and reds depends on length of time in **contact** with dark purple grape skins.
- Color saturation tends to match flavor **intensity**, and may indicate **oakiness** or **astringency**.
- The tongue can detect just **six taste sensations**, of which only **sweetness** and **sourness** are relevant to wine tasting.
- In the wine world, the term **"dry"** means "not sweet" rather than "not wet."
- The naturally **high acidity** of fresh grapes makes wine more tart than most beverages. This helps it flatter food, refresh the palate, and age gracefully.
- **"Fruit"** is the term given to all olfactory scents and flavors that come from the grapes used to make wine.
- **"Oak"** describes a narrow range of scents and flavors derived from a wine's contact with wood.
- **"Mouthfeel"** describes the tactile sensations we discern with tongue, palate, lips, and gums.
- The lower a wine's **alcoholic strength**, the lighter it feels on the palate; the higher its strength, the heavier it seems.
- **Tannin** is an astringent phenolic compound present in the skins, seeds, and stems of grapes. Found only in **red wines**, it dries out the mouth after wine is swallowed.

BROWSING AND BUYING

TRICKS OF THE TRADE Wine is harder to shop for than almost any other product. The options in stores and restaurants can paralyze even the most savvy shopper. However, picking up a few strategies wine professionals rely on can help banish doubts and build wine confidence. Labels may not tell us how wine will taste, but there's much to be learned between the lines. Wine packaging speaks volumes about what's inside, and three important numbers can simplify the shopping experience. A few restaurant rules of thumb make it easier to benefit from expert advice without breaking the bank.

LOOK AT THE PACKAGING

There is quite a lot we can learn about a wine without even reading the label—colors, fonts, and artwork speak volumes. Since wine-label terms and regulations can seem so disorienting to consumers, vintners often try to communicate style without words, through their package design.

SUBLIMINAL LABEL MESSAGES

Packaging design reflects the vintner's style and philosophy and often hints at how the wine will taste.

JUDGING WITHOUT TASTING

In the modern self-serve retail environment, producers have every incentive to make sure their packaging evokes the wine's sensory profile and appeals to the appropriate audience.

FUN-LOVING AND MODERN

Splashy colors, modern designs, and lighthearted names suggest ripe wines with dessert-like aromas, meant for immediate consumption.

SOPHISTICATED CLASSIC

Sedate colors, traditional designs, and family names suggest more food-friendliness, tending toward leaner, drier wines with noticeable acidity.

BRIGHT EQUALS LIGHT

Among whites, clear glass and blue-green colors suggest lighter-bodied, unoaked wines. Darker glass and fall colors hint at a fuller body or oakier flavor.

IDENTIFY A WINE BY ITS BOTTLE

Some wine regions use distinctive bottle shapes that have come to form part of their identity—an indicator of the style found within.

BURGUNDY

So-called Burgundy bottles with sloping shoulders are most often used for Chardonnay, Pinot Noir, Syrah, and Rhône-style wines.

BORDEAUX

Square-shouldered Bordeaux-style bottles are used for most other styles, particularly Cabernet Sauvignon, Merlot, and Sauvignon Blanc.

ALSACE

Tall, fluted Alsace-style bottles are associated with white wines of German heritage, such as Riesling and Gewurztraminer, which may be sweet or dry.

PACKAGING INNOVATIONS

Some wines can age for decades, so wineries have been cautious in adopting modern packaging innovations. However, bottles with corks are not the most efficient containers, and today's vintners are exploring new options.

ALTERNATIVE CLOSURES—LIKE SCREW CAPS

Natural corks are still the norm for fine wine, but they can cause detectable flaws in a small percentage of the wines they are supposed to protect. Screw caps, offering more consistency and a lower failure rate, are becoming more commonplace, particularly for wines designed to be drunk young.

ALTERNATIVE CONTAINERS—LIKE BOXES

Bottles are no longer the only option for preserving wine. Improvements in food-packaging technology mean boxes, cartons, and cans are now viable containers. All of these can reduce shipping weight and protect wine from damaging light. Larger "bag-in-box" formats have the added advantage of extending wine's shelf life dramatically once they're "tapped"; the shrinking bag inside protects the wine from air, so it can taste fresh for up to six weeks.

SHOPPING BY NUMBERS

Even when we know what style of wine we want—an Italian red or a California white—making a final decision can be complicated by the sheer number of choices. When faced with a proliferation of unfamiliar options or similar wines, professionals often compare three key numbers, each of which provides clues to how the contenders will taste.

CHECK THE VINTAGE

The year printed on the label is far more useful for learning how old a wine is than for whether it was made in a "good year" or "bad year."

The younger a wine is, the more fruit-driven it is likely to be. Young wines feature fresh-fruit flavors, and the youngest are often unoaked. They are also likely to be lighter-bodied, more refreshing, and more affordable.

Wines that are more than two years old are typically premium wines refined by aging, often in oak barrels. Mature wines feature flavors that are less vivid and fresh but more complex and multilayered, often with a distinctly toasty oak accent. Such wines are typically fuller-bodied, more opulent, and more expensive.

2007 2008 2009 2010 2011 2012 2013 2014 2015 2016 2017 2018 2019

CONSIDER THE ALCOHOL

A wine's strength (its alcohol by volume, or ABV) can be used as a rough indicator of many important style parameters, especially body and flavor intensity. This idea is explored in greater detail in the next section, but even a superficial overview can provide relevant insights when deciding on a wine to buy or drink. As a broad generalization, you can reasonably expect certain wine traits to vary proportionally with alcohol content.

UNDER 13.5%	13.5–14%	OVER 14%
THE LOWER THE ALCOHOL, THE MORE LIKELY A WINE WILL BE …	STANDARD MID-WEIGHT WINES	THE HIGHER THE ALCOHOL, THE MORE LIKELY A WINE WILL BE …

UNDER 13.5% — THE LOWER THE ALCOHOL, THE MORE LIKELY A WINE WILL BE …

- lighter, or more sheer in texture
- paler in color
- less intense in fruit and oak flavor
- higher in refreshing acidity
- younger and more fresh tasting

13.5–14% — STANDARD MID-WEIGHT WINES

OVER 14% — THE HIGHER THE ALCOHOL, THE MORE LIKELY A WINE WILL BE …

- heavier, or thicker in texture
- deeper in color
- more intense in fruit and oak flavor
- lower in refreshing acidity
- aged and richer tasting

LOOK AT THE PRICE

The most important number in buying decisions is always price. Many drinkers feel they need to spend more than they'd like on wine to get a "decent" bottle. Wine professionals know that the biggest jumps in quality are found among the most affordable wines and that quality is not always the biggest concern.

In wine, as with all consumer goods, higher prices correlate to higher quality. Making wine with better materials and craftsmanship yields more desirable traits, such as aesthetics, individuality, and durability, just as it would for making shoes or cars. However, there is a point of diminishing returns. (See the next page for more insight into shopping by price.)

(See the next page for more insight into shopping by price.)

IT COSTS HOW MUCH?!

We're used to paying more for quality in all consumer goods, but wine's lofty price tags can appear startling. In some instances, they seem to defy logic. A bargain Rioja might cost $10, while a great estate's top *gran reserva* could easily command $100. At auction, aged bottles might sell for $1,000 or more. Why? Finer wines are more expensive to make, but making better wines also means making fewer bottles at much greater expense, sometimes aging them for a decade or more before they're sold. The rest comes down to supply and demand.

WINES THAT RETAIL AT $100 ARE CERTAINLY HIGHER IN QUALITY THAN WINES THAT COST $10, BUT THIS DOES NOT MEAN THEY ARE 10 TIMES BETTER OR THAT YOU WILL LIKE THEM 10 TIMES AS MUCH.

BEATING THE BUDGET

Wine has a reputation as a luxury product, but drinking well doesn't mean you have to break the bank. Wine professionals don't drink expensive wines every day, because they know how to get the most bang for the buck. Consider some of these insider tips next time you're shopping for wine.

TRY A NEW PRICE POINT

You don't have to spend a lot more per bottle to drink significantly better wine. For example, let's say the popular price for everyday wine in your area is $10 per bottle. Wines priced just above this, in the $11 to $15 range, will typically offer the best mileage for the dollar in terms of quality for the price. If they weren't worth more than the cheaper bottles in some way, they would be unable to command the higher price. As prices continue to rise from $20 to $40 and beyond, we can expect the wines to continue to improve, but we will have to pay a greater premium for each step up in quality.

Cheap, but hit-and-miss

Safer bets, offering best value mileage

Superior versions for the quality-conscious

BARGAIN   **AFFORDABLE**   **PREMIUM**

BE AN EXPLORER

Famous wine regions such as Napa Valley will command higher prices than obscure ones like Paso Robles, and the same is true of legendary grapes, trusted brands, and fancy package designs. These demand-driving factors do correlate with increased odds of better wine quality, if only by signaling that the vintner has the resources to pursue ambitious goals. However, there are many great wines that can be found by those willing to go off the beaten path. Although trying unfamiliar wines can seem risky, people often overestimate the danger. There's very little truly bad wine out there, and the only way to find out what you like is to jump in and explore.

TAKE REVIEWS WITH A GRAIN OF SALT

Wine reviews can seem like lifesavers to the novice, but shopping according to point scores often results in overpaying. Since magazine scores don't take prices into account, there's a strong correlation between higher scores and wines that are more expensive to produce. Good scores also increase demand, providing an incentive for wineries, brokers, and retailers to raise prices on top-ranked wines—or at least to resist implementing discount strategies.

LOAD UP YOUR CART?

Most retailers offer discounts to those shoppers who buy their wine in bulk, but this is not the only way for you to benefit from "volume pricing." Look for the items that your retailer purchases in greatest quantity. Floor stacks and feature bins show confidence in the product and likely have a lower markup than less well-stocked items.

Decadent limited-production splurges

Rare specialties for when price is no object

LUXURY

COLLECTIBLE

READING NEW WORLD LABELS

All wine labels list the vintner or brand identity along with their formal appellation, or region of origin. The most famous wine appellations tend to be small wine regions. However, political districts also qualify as wine appellations in many countries, ranging from large provinces to smaller counties and municipalities. Labels on wines from the New World—the Americas and the southern hemisphere—tend to be easier to understand, so that is where we will begin …

GRAPES UP FRONT

Most modern wine labels advertise the name of the grape variety from which the wines are made, as well as their vintage date; however, these details are not required by law.

WHAT'S ON THE LABEL?

I. WINE APPELLATION OR REGION OF ORIGIN—MANDATORY
A formal region-of-origin statement is required for all wines, indicating where the grapes were grown, which is not necessarily where the wine was made.

2. BRAND NAME OR WINE PRODUCER—MANDATORY
Wines are most often sold under the name of the winery but may also be sold under a proprietary brand name.

3. VINTAGE DATE—OPTIONAL
This specifies the year when the grapes were harvested.

4. GRAPE VARIETY—OPTIONAL
Wines are typically labeled by the type of grape used, in which case they must contain at least 75 percent of the specified variety.

5. FINE PRINT
Legal requirements vary by country, but all wine bottles must indicate (on front or back label): bottle volume, alcohol content, and country of origin. The company or facility that made the wine must also be formally identified, along with its location.

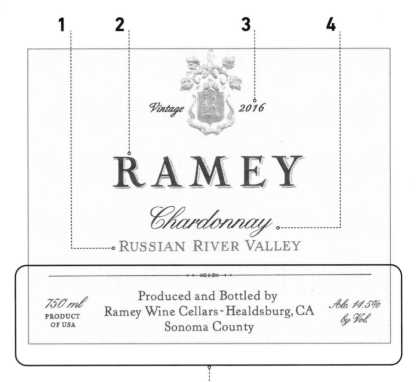

1 **2** **3** **4**

Vintage *2016*

RAMEY

Chardonnay

RUSSIAN RIVER VALLEY

750 ml
PRODUCT
OF USA

Produced and Bottled by
Ramey Wine Cellars ~ Healdsburg, CA
Sonoma County

Alc. 14.5%
by Vol.

5

KNOWING WHAT'S WHAT

Whether they make wine in Europe or the New World, vintners often make more than one wine from the same grape and region—a basic version and a premium version, for example, or a sweet version and a dry version. To distinguish these from one another, they are given *cuvée* names that provide additional specificity on the label. (*Cuvée*, which is French for "vat," loosely means batch or blend in this context.) The optional extras that may appear in a wine's *cuvée* name include the following.

1. VINEYARD SITE NAMES
such as Dutton Ranch or Ornellaia.

2. QUALITY TERMS OR CERTIFICATIONS
such as "reserve" or "organic" (some regulated, some not).

3. WINE-SPECIFIC PROPRIETARY NAMES
such as Penfolds Grange or Insignia by Joseph Phelps.

4. MULTI-WINE BRAND NAMES
such as Rothschild's Mouton Cadet or Coppola's Diamond Label.

5. STYLE REFERENCES
such as "sparkling" or "late harvest."

1

5

3

2

4

READING OLD WORLD LABELS

Following centuries of tradition, top European wines are named according to their regions—Chablis and Chianti, for example—not their grapes. As a result, the European Union (EU) regulates wine-label statements in its own way. In "Old World" Europe, appellations are the organizing principle for wide-ranging quality standards, as well as their enforcement mechanism; elsewhere, appellations simply certify where a wine's fruit was grown.

REGION OVER GRAPES

Vintners within top European appellations may grow only specified grapes. They also must conform to strict winemaking requirements in order to earn the right to use a place name on the label. Appellations formalize local traditions and are often organized as hierarchies, with smaller, premium appellations carving out the best sites within larger, less prestigious ones.

There are a few regions in Europe where naming grapes is traditional, such as Germany and northern Italy. Many up-and-coming appellations elsewhere also list grapes to appeal to international markets. However, in the French system that served as a model for EU wine law, a wine's region is its distinguishing feature and point of pride, not its grape. For example, the French Chablis below is made entirely with Chardonnay by law, but the label makes no reference to grape variety at all. This pattern is most often seen in the most traditional and ambitious wines.

HIDE THE GRAPE

Wines from Europe's prestigious regions have been named by their source for centuries—whether it's a large region or a tiny village. Their dominant grape is often not mentioned on the label.

EUROPEAN WINE APPELLATION

PRIDE OF PLACE

Region-of-origin statements are regulated differently in Europe and classified according to their quality standards. Premium appellations are typically prominent on labels, and the formal legalese is in small print immediately adjacent.

GRAND VIN DE BOURGOGNE

J. MOREAU & FILS

MAISON FONDÉE EN 1814

CHABLIS

APPELLATION CHABLIS CONTRÔLÉE

CHABLIS · FRANCE

RANKING OF QUALITY

Some European wine regions add their own special terms on labels as quality indicators. Such designations are tightly controlled, often recognizing multiple ascending ranks. Each region uses its own system, which can be based on one or more wine quality factors.

1

2

3

1. SUPERIOR VINEYARD LAND
In some French regions, the best vineyards or wine estates are granted elite *grand cru* or next-best *premier cru* status. In Italy, *classico* signifies a superior subdistrict.

2. EXTENDED BARREL MATURATION
In Italy and Spain, legally defined terms such as *riserva* and *crianza* respectively are used to distinguish wines that spend extra time aging in oak barrels.

3. DEGREES OF GRAPE RIPENESS
In Germany and Austria, full ripeness is not guaranteed, so a complex system of label terms like *Kabinett* and *Spätlese* is used to rank wines by the sugar content of their grapes at harvest.

PROS AND CONS

The upside of European wine labels is that they often include lots of information that is meaningful and subject to regulatory control. However, most is only intelligible if you're already an expert. For example, the formal name of this Champagne includes six different categories of information.

SWEETNESS LEVEL

STYLE CATEGORY

APPELLATION

PRODUCER

VILLAGE NAME AND QUALITY RANK

BUYING IN RESTAURANTS

Restaurants offer an opportunity to sample multiple wines, both alone and with food, from a selection carefully chosen to complement their cuisine (see the "Matching Wine and Food" chapter for food-and-wine matching tips). However, restaurants can also present challenges for wine lovers—from poor wine service to cynically high price markups.

CAN YOU TRUST THE VENUE?

Restaurants and bars can offer outstanding wine experiences tailor-made for sharing with friends and family—but only if the business cares about its wines and its customers. Take a quick look around. If the bar or restaurant takes pride in its wine service, you'll see the signs before you order. If their wine list looks like an afterthought, or if they would clearly prefer to sell you beer or cocktails, proceed with caution.

GOOD SIGNS

- Wine glasses already set on tables
- Large wine glasses filled less than halfway
- Clean, well-organized wine lists
- Many wines offered by the glass
- Extra wine details on the list, such as style descriptions

BAD SIGNS

- No wine glasses visible
- Small wine glasses filled near the top
- Scruffy, poorly organized wine lists
- Few wines offered by the glass
- Incomplete wine listings—with no vintage dates, for example

SEEK OUT ADVICE

Not all restaurants have skilled, wine-savvy staff available to advise their guests, but when they do you may as well take advantage. Whether it's a certified sommelier or a well-trained bartender, those who work most closely with their wines are great resources. Don't simply ask for their suggestions, unless you truly want to try the wines they like the most. If what you want is a wine closer to your personal tastes, provide them with more to go on. For example, one great strategy is to say, "I love Pinot Grigio but would like to try something different tonight. What can you recommend?"

STAY IN CONTROL!

It's always in a restaurant's interests to boost wine sales, so sommeliers and servers will often favor premium wines and extra bottles unless otherwise directed. Many diners feel intimidated when ordering wine and go along with wine suggestions for fear of coming across as a stingy host. However, as the paying guest, remember that you are always in charge.

NO ONE CAN FORCE YOU TO SPEND MORE IN A RESTAURANT THAN YOU WISH UNLESS YOU CEDE CONTROL.

PROVIDE A PRICE TARGET WHEN ASKING FOR RECOMMENDATIONS

Discreetly signal your budget by pointing to a price on the wine list. Without guidance, the server won't know whether you want to scrimp or splurge. Specifying a price point directs them to flavor options instead of quality levels.

DON'T ORDER A BOTTLE WHEN A GLASS WILL DO

Wines by the glass and half-bottles are ideal for providing a palate-cleansing aperitif while reading the menu or for a final drink to finish off the main course or cheese plate. Bottles may feature the lowest margins, but they're only the best deal if you finish them.

DON'T APPROVE IN ADVANCE UNLIMITED BOTTLES FOR LARGE PARTIES

It's tempting to tell servers to keep the wine flowing so you can relax and enjoy yourself. But doing so allows the staff to decide how much you'll spend. Expect to need one bottle for every two wine drinkers at dinner; less at lunches and brunches. If approving each new bottle is intrusive, authorize a set number.

CHAPTER CHECKLIST

Here is a recap of some of the most important points you've learned in this chapter.

- The information printed on **wine labels** can be disorienting to the general public, so many producers try to communicate their **wine style** through their **packaging**.

- Bottles with **corks**, though traditional, are not the most efficient wine packages. **Screw caps** are now often used instead of corks, and **boxes** are increasingly replacing bottles for everyday wines.

- When shopping for wine, three important **numbers** can serve as useful guides: vintage, level of alcohol, and price.

- Not even wine professionals drink expensive wine every day. Learn how to get the **best value** even on a budget, and don't feel under any pressure to **overspend**.

- **Wine labels** rarely reference **flavor**. Most give details that are more helpful for professionals than casual wine drinkers, such as the **type of grapes** used to make the wine and where they were grown.

- The most useful **label-reading skills** are to recognize distinctions between the main types of label **statements**, and between grape-centric labels and European region-centric labels.

- When dining out, consider whether you can put your trust in the restaurant based on the amount of effort and pride they put into their **wine service**.

- At a restaurant where a **sommelier** or wine-savvy bartender is present, take full advantage and seek their **advice**.

- Be clear to your waiter about how much you wish to spend on wine and **don't be bullied** into spending more, whether by ordering a more expensive bottle than you would choose or by ordering more wine than is necessary.

POURING
AND
STORING

WHAT TO DRINK WHEN AND HOW Wine is an indulgence we can enjoy at home and share with friends and family. However, unlike beer and spirits, how to serve it and store it isn't always clear cut. Wine behaves differently at different temperatures, affecting how it tastes and how well it keeps. Understanding how and why makes it easier to get the most enjoyment from your wine purchases and to maximize their shelf life, even after the bottles are opened. Knowing how much to pour into what type of glass, or how many bottles you'll need, makes home entertaining with wine a breeze.

ENTERTAINING WITH WINE

Whether you're dining out or entertaining at home, it's helpful to know how much wine you'll need and how to serve it. Here are some general guidelines.

HOW MUCH WINE IS IN A GLASS?

Internationally, an average glass of wine is 5 fl oz (148 ml), and there are about five such servings in a standard bottle. However, certain situations dictate pouring roughly half this amount:

- When a bottle is shared at the table, especially in fine-dining restaurants, or for multicourse meals that include more than one wine
- When the wine's primary purpose is to raise a toast
- When the wine is extremely sweet or extremely strong, as is the case with dessert wines and fortified wines
- At wine tastings where multiple wines are being sampled

PORTION SIZES

FULL-GLASS POUR (5 FL OZ, OR 148 ML)
This is the correct portion for:
- cocktail parties
- wine before dinner
- casual meals (single course, single wine)
- wines by the glass in restaurants

HALF-GLASS POUR (2½ FL OZ, OR 74 ML)
This is the correct portion for:
- toasts and wine tastings
- after-dinner wines
- fancy meals (multiple courses, multiple wines)
- wines by the bottle in restaurants

HOW MUCH WINE DO YOU NEED FOR A PARTY?

Having 1 bottle per person on hand is wise when entertaining, to avoid running short; but average consumption is typically about half this amount.

- For social events and receptions: 1½ glasses per person for the first hour, then 1 glass per person per hour thereafter.

- For dining out: 2½ glasses per person, or 1 bottle for every 2 guests.

- For dinner parties: 2 bottles of reception wine, then 1 bottle per course for every 6–8 drinkers.

- For wine tastings (with 6–10 wines): 1 bottle of each wine for every 10–12 guests.

HOW MUCH WINE IS IN A BOTTLE OR A BOX?

Wine bottles look and feel substantial, so people often overestimate how much wine they contain.

Standard bottles hold 750 ml, or 3 cups, of wine. The most common size for premium box wines is 3 liters; this is the equivalent of four standard bottles, but it occupies quite a bit less space.

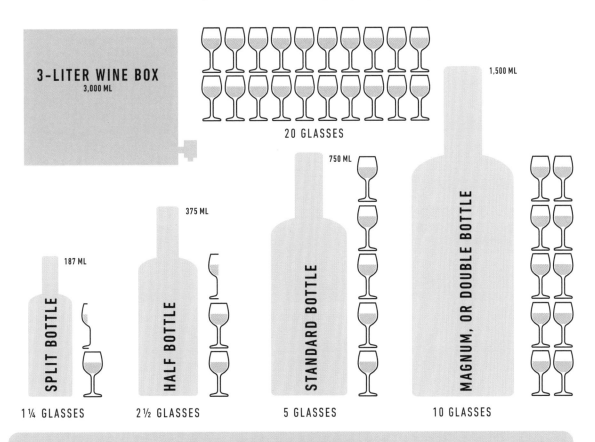

3-LITER WINE BOX
3,000 ML

20 GLASSES

1,500 ML

750 ML

375 ML

187 ML

SPLIT BOTTLE

HALF BOTTLE

STANDARD BOTTLE

MAGNUM, OR DOUBLE BOTTLE

1 ¼ GLASSES   2 ½ GLASSES   5 GLASSES   10 GLASSES

WHAT TO OFFER WHEN

It can be daunting trying to figure out just what people expect when you are hosting an event. Here are some points to remember.

- Traditionally, the reception drink—served to guests on arrival—is sparkling wine, because it is light and piques the appetite.

- Wine preferences (white or red) tend to be roughly 50/50, so guests usually expect both types to be available. Choose crowd-pleasers: Mid-weight wines have the broadest appeal.

- Offer a lightly sweet option at larger events, especially those that attract multiple generations of guests. Many

people never acquire a taste for dry wines and prefer sweeter styles.

- Daytime events call for lighter wines. Look for younger, lower-alcohol styles and expect less total consumption than for evening affairs.

- For multicourse, multi-wine dinners:

  1. Serve sparkling wine as an aperitif and/or with first courses.

  2. Serve white wines before reds—lighter before heavier if more than one.

  3. Finish with dessert wines or fortified wines.

KNOW YOUR GLASSES

Wine tastes good enough to be enjoyed from any vessel, even sipped straight from the bottle, but it is usually served in specialized wine glasses. Much as darkened theaters flatter movies by focusing our attention on the bright screen, stemmed glasses with large bowls flatter wines by focusing our attention on their aromatics.

WINE-GLASS ANATOMY

Most glasses and cups are designed for efficiency and convenience, like tumblers that get filled to capacity. Wine glasses are not; they are designed to please your nose. Every feature is oriented around swirling, sniffing, and maintaining wine's temperature—all to enhance wine's scent. There are all sorts of glasses, some made for specific styles of wines. But a single multipurpose wine glass is all you need to enjoy everything from Prosecco to Port.

HEADSPACE
This part of a wine glass stays empty for two reasons: to allow the wine to be vigorously swirled without spilling, increasing its surface area and its rate of evaporation; and to retain and concentrate the resulting aroma compounds.

WINE
Wine glasses are designed to hold 5–6 fl oz (15–18 cl) of wine. As a rule, they should not be filled more than halfway, or above the widest point of the bowl to maximize the impact of the wine's aroma. As a result, 10–12-fl oz (30–35-cl) glasses are ideal.

STEM
The stem of a wine glass serves as a handle, designed to keep your fingers off the bowl. Body heat is easily transferred, and wine's flavor changes greatly with even small shifts in temperature. The stems should comfortably accommodate the width of your hand at the knuckles.

BASE
Well, you might need to put the glass down. Possibly.

RIM
Wine glasses are most fragile at the lip, so practical glasses often feature sturdy rolled rims. Deluxe glasses have thinner cut rims that can feel amazingly sheer on the lips but make them easy to break.

BOWL
Wine glasses should feature large bowls that narrow at the top to concentrate wine's aromas. The most versatile are widest near the bottom and have a total capacity of roughly 10–12 fl oz (30–35 cl).

THE SUM OF ITS PARTS
Wine glasses are unusual in that their component parts are designed more for smelling than for drinking.

CHOOSE THE RIGHT SIZE

White wines are often served in smaller glasses than reds because they smell milder; their pleasing scent can seem weak when there is too much headspace. Conversely, reds can smell too strong in too small a glass, since they contain more aroma compounds. When using one all-purpose glass, compensate by filling it higher for whites and lower for reds.

In casual cafés, wine often comes in tiny 6–8-fl oz (18–24-cl) "Paris goblets" filled near the rim, limiting your ability to smell it properly. In fine-dining restaurants, expensive wines may be served in huge crystal glasses that could accommodate a whole 75-cl bottle. Specialized stemware for particular grape varieties or styles enhance appreciation slightly, but are neither practical nor necessary.

WHITE WINE STYLE   RED WINE STYLE   CAFÉ STYLE   LUXURY STYLE

A GLASS APART

Two styles of wine benefit from different glasses entirely: sparkling wines and highly concentrated wines. Bubbly wines lose carbonation faster with a larger surface area, so they are served in tall, thin flutes. High-sugar dessert wines and high-alcohol fortified wines are served in half-size 2½–3-fl oz (7–9-cl) portions. Their aromatics are typically strong enough to enjoy in a standard wine glass, but they look nicer served in smaller glasses.

SHERRY GLASS AND CHAMPAGNE FLUTE

WINE AND TEMPERATURE

Wine's flavor is dependent on volatile components that are temperature-sensitive, so we try to serve it at the most flattering temperature and maintain that ideal by pouring small amounts, keeping bottles on ice, and holding the glass by its stem. We have easy access to room temperature and refrigerator temperature, but most people prefer their wines somewhere in between: a little cooler than the room for reds and a little warmer than the fridge for everything else.

WHY DON'T WE CHILL REDS?

The only wines that aren't routinely served cold are red wines, because the tannins and other grape-skin compounds that give them color seem more astringent and bitter at low temperatures.

YOUR WINE, YOUR RULES!

As with all things, individual preferences in wine temperature can vary widely, and the most important thing is to drink your wine the way you like it. If that means icing your reds or warming your whites, ignore the snobs and go for it!

Remember, though, that if your wine seems a little too bland, it may just need to warm up a little; and if it seems a little too boozy or lacks refreshment, it may just need a little chill.

DISCOVER HOW TEMPERATURE CHANGES WINE FLAVOR

Open two bottles of wine—one chilled white and one red at room temperature.

- Pour two glasses of each wine, and put one of each in the refrigerator.
- After 10 minutes, taste each pair side by side, looking for the differences in taste.

Both wines from the fridge will seem less aromatic and less flavorful than the same wine in the warmer glass—but also more refreshing and less boozy. The colder red's tannin will seem harsher and more raspy, too. The reverse will be true of the warmer pair, seeming more flavorful and fragrant, as well as more alcoholic and less refreshing. And the red's tannins will feel smoother.

Most people will find the cold red to be unappealing and the white better once it has been out of the fridge a little while.

OPTIMUM TEMPERATURES

It is perfectly acceptable for you to drink wine at whatever temperature you like—even chilling your reds, for example. However, you may wish to follow standard practice when you have guests.

SERVING ADVICE FOR REDS

- Red wines taste best at 60–70°F (15–21°C).
- Serve lighter reds colder and heavier reds warmer.
- Keep reds at room temperature, then chill briefly immediately before serving: 5–15 minutes in the refrigerator. For example, chill lighter French Beaujolais for 15 minutes before serving, but give heavier Argentinian Malbec just 5 minutes.
- Port wines are an exception to this rule: Although they are red and strong, they taste more balanced when chilled as for whites.

SERVING ADVICE FOR OTHER WINES

- White, rosé, sparkling, fortified, and dessert wines taste best at 40–50°F (4–10°C).
- Serve lighter wines colder and heavier wines warmer.
- Keep these wines refrigerated, "warming" them briefly immediately before serving: 5–15 minutes at room temperature. For example, let heavier Australian Chardonnay shed its chill on the table for 15 minutes before serving, but give lighter Italian Pinot Grigio just 5 minutes to warm. Serve the lightest wines, like Spanish Cava, straight from the fridge.
- Sweet dessert wines are an exception to this rule: They taste best quite cold, regardless of their alcoholic strength.

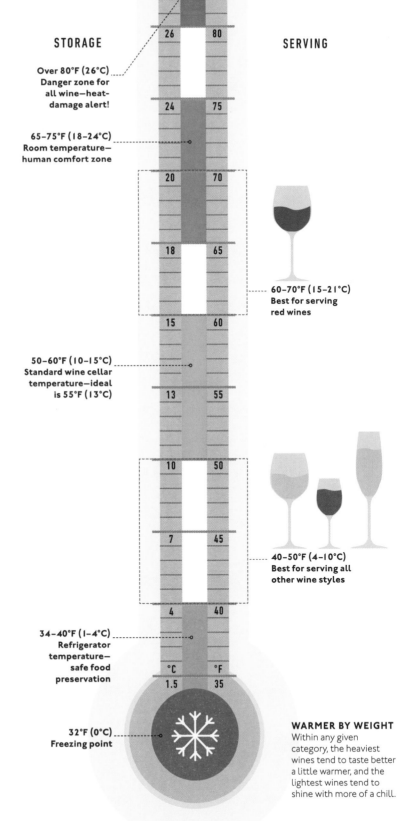

STORAGE

SERVING

Over 80°F (26°C) Danger zone for all wine—heat-damage alert!

65–75°F (18–24°C) Room temperature—human comfort zone

60–70°F (15–21°C) Best for serving red wines

50–60°F (10–15°C) Standard wine cellar temperature—ideal is 55°F (13°C)

40–50°F (4–10°C) Best for serving all other wine styles

34–40°F (1–4°C) Refrigerator temperature—safe food preservation

32°F (0°C) Freezing point

WARMER BY WEIGHT
Within any given category, the heaviest wines tend to taste better a little warmer, and the lightest wines tend to shine with more of a chill.

PRESERVING AND DECANTING WINE

The fun begins once your wine is opened. Historically, serving vessels were used purely for practicality's sake, but modern decanters are designed to enhance wine's aesthetic appeal. Well-made wines that are not yet mature can briefly improve with aeration, while older wines can be poured off their sediment. Unfortunately, preserving unfinished wines for later enjoyment is tricky.

INSPECT YOUR GADGET

Wine preservation accessories that rely on inert gas can be reasonably effective. However, vacuum-based devices are not. They do slow oxidation as advertised but at the expense of speeding up flavor loss.

OXYGEN: FRIEND AND FOE

Wine is not inert; its flavors change slowly as it matures in barrels and continue to change as it ages in bottles. Both of these positive processes are driven by wine's interaction with the oxygen present in air, but at a snail's pace. When a cork is pulled and wine is poured, the rush of oxygen-rich air triggers rapid changes that alter the taste of the wine in the glass even as we drink it. In high-quality wines that are not yet mature, the wine may even taste better for a day or two once opened. However, lesser wines and older wines will succumb quickly to oxidative decline, browning in color and losing freshness of flavor.

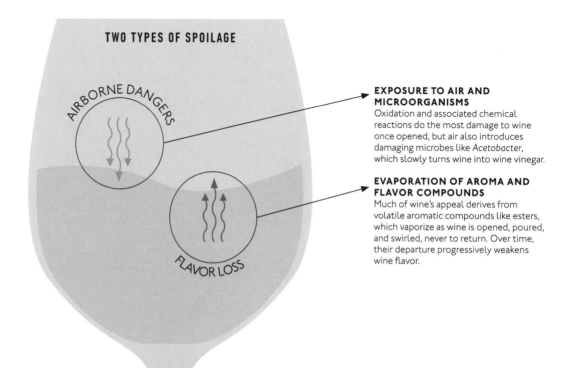

TWO TYPES OF SPOILAGE

AIRBORNE DANGERS

FLAVOR LOSS

EXPOSURE TO AIR AND MICROORGANISMS
Oxidation and associated chemical reactions do the most damage to wine once opened, but air also introduces damaging microbes like *Acetobacter*, which slowly turns wine into wine vinegar.

EVAPORATION OF AROMA AND FLAVOR COMPOUNDS
Much of wine's appeal derives from volatile aromatic compounds like esters, which vaporize as wine is opened, poured, and swirled, never to return. Over time, their departure progressively weakens wine flavor.

DECANTERS AND CARAFES

Serving vessels for wine have become more specialized now that modern wine bottles fulfill their original function.

EXCESS CAPACITY
Many modern decanters are large enough to hold two or more bottles, but are designed to give a single bottle ample space for swirling and plenty of air.

WIDE BASE
Modern decanters provide a large surface area to maximize wine aeration, which benefits younger wines but poses a danger for mature wines that are at or past their peak.

PORTION CONTROL
In restaurants, metered carafes are useful for doling out consistent volumes of wine from larger bottles, boxes, or casks.

NARROW BASE
A small footprint conserves space on tables and in cabinets, and helps make carafes easy to pour as well.

LUXURY DECANTER

RESTAURANT-STYLE CARAFE

WHY DECANT WINE?

Decanters have long been used to make wine look pretty, but improving flavor by decanting is relatively new. For most of the 20th century, decanting was meant to separate mature red wines from the sediments they develop after 10 or more years of aging. However, since fine wines are increasingly consumed earlier in their evolution, decanting is more often practiced to help younger wines taste more mature by helping them "breathe," or aerate. Aeration is only beneficial for high-quality wines that have yet to reach their peak, so premium red wines and dessert wines are decanted more often than other styles. How long a wine should breathe for maximum benefit depends on the resilience of the wine in question, and is a topic subject to spirited debate even among experts. In practice, though, very few wines are concentrated enough to taste appreciably better after more than an hour or two.

NEED TO PRESERVE OPEN WINE? FREEZE IT!

It may sound sacrilegious, but the simplest and most effective way to preserve opened wine is to freeze it.

Freezing cannot save every shred of flavor in fresh wine—nothing can—but it is shockingly efficient at retaining the vast majority of its appeal. Best of all, no other preservation method has its staying power. A frozen wine will taste the same whether it is thawed after a day, a week, a month, or a year.

Bear in mind that frozen wines will throw a harmless sediment once thawed. As with other preservation methods, young, bold-tasting wines are more resilient than mature or delicate-tasting wines. Thawing a frozen bottle takes a few hours on the counter, but can be sped up under warm running water. A few limitations do apply, though—see right.

NEVER FREEZE UNOPENED BOTTLES

DON'T FREEZE MATURE OR SPARKLING WINES

1959

DON'T LET THE WINE TOUCH THE CORK

WINE AND AGING

Any wines that will be drunk within a couple of months can be stored in the refrigerator or at room temperature, but long-term storage requires more wine-friendly conditions—specifically a cool temperature and protection from light. Wine is a perishable product that changes over time in a slow march toward spoilage. It may take them longer to go bad than fresh grapes, but eventually all wines will succumb to oxidation, turn brown, and die.

NOT ALL WINES AGE WELL

Contrary to popular belief, most wines do not improve with age. Many can be kept quite a while before they deteriorate, but their fresh-fruit qualities will fade, and very few wines are concentrated enough to develop new flavors and smells to replace them over time.

The distinctive, complex scents of mature wines arise from chemical reactions that take place in the bottle between components like phenols and esters. Wines that don't have enough such "stuffing" from the start will tire and lose intensity rather than becoming more interesting with age. And wines that can't fend off the onslaught of oxidation with preservative components like antioxidant tannins or high levels of acidity don't stand a chance. Only uncommonly concentrated wines reward patient cellaring, and these tend to be expensive styles.

RATIO OF WINE AGEABILITY

OVER 90% OF ALL WINES ARE DESIGNED TO TASTE BEST IMMEDIATELY.

THE FACTS:

- This is especially true of white, rosé, and sparkling wines.
- Any necessary aging is done at the winery before release.
- Well-made wines won't decline immediately, but rarely improve.
- Value-oriented wines tire fastest.
- Pink rosé wines and nouveau-style reds are the least stable and should be drunk within 6 months.

LESS THAN 10% OF WINES ARE DESIGNED TO TASTE BETTER AFTER 5 YEARS.

THE FACTS:

- They must be high enough in concentration to develop new flavors.
- Most in this bracket are red wines or dessert wines, since high levels of tannin and sugar can act as preservatives.
- Typically, these are premium and luxury wines, usually made in a traditional style.

LESS THAN 1% OF WINES ARE DESIGNED TO TASTE BETTER AFTER 10 YEARS.

THE FACTS:

- The finest high-tannin reds may not taste pleasant when very young, because they tend to need a few years to "soften."
- These types of wines are usually exceptional collector's wines, both rare and expensive.

Decades ago, it was easier to predict which wines needed to be laid down: you just needed to remember a few grapes and styles! Nowadays, the market prefers wines that are ready to drink, and most vintners do whatever it takes to comply. But making wines to taste better sooner means they will also decline faster. The only way to be sure a wine will improve with age is to open a bottle and see what happens. Pour yourself a glass or two, and leave the rest of the bottle sitting out on the counter. If the wine tastes better the next day, it can improve with age; and the longer it continues to taste pleasing, the more likely it is to reward a few years of cellaring. But if the wine's flavor flattens or becomes less pleasant, drink it over the next few months.

HOW TO STORE WINE

For wine, the ideal storage conditions are like those found in a natural basement: dark and damp, still and chilly. A temperature range of 50–60°F (10–15°C) is ideal for wine's development, but since few people can afford a cellar or wine fridge, it may be necessary to compromise. Pantries and closets that don't get too hot or dry can work for laying wines away—ideally, sealed in boxes and laid on their sides on the floor. Natural corks shrink when they dry out, and this allows damaging air in. Storing bottles horizontally keeps them wet and plump, maintaining their protective seal.

CHAPTER CHECKLIST

Here is a recap of some of the most important points you've learned in this chapter.

- An **average glass of wine** is 5 fl oz (148 ml), so a standard-size bottle holds about **five glasses**.

- Standard bottles hold **750 ml**. The most common size for box wines is **3 liters,** the equivalent of four standard bottles.

- Wine is delicious enough to drink from any vessel, but it tastes better served in specialized **wine glasses**.

- Wine glasses are designed to please your nose, their shape based on **swirling, sniffing, and maintaining** wine's temperature—all to enhance the scent of the wine.

- There are two reasons for **decanting**: to remove **sediment** in older reds, and to **aerate** and mellow younger wines.

- The flavor of wine depends on **volatile components** that are sensitive to temperature. This is why it is best to serve it at the most **flattering temperature** and maintain that while drinking, too.

- **Individual preferences** in wine temperature can vary greatly. Ultimately, you should drink your wine the way you like it.

- The only wines not routinely served cold are reds, because their tannins and other grape-skin compounds tend to seem more **astringent and bitter** at low temperatures.

- **Red wines** taste best at 60–70°F (15–21°C). All other wines—whether **white, rosé, sparkling, fortified, or dessert wines**—taste best at 40–50°F (4–10°C).

- Many wines can be kept quite a while before they **deteriorate**, but very few are **concentrated** enough to develop new flavors and smells to replace those lost over time.

NAVIGATING
WINE
BY
STYLE

THE HUGE NUMBER OF WINE options and impenetrable information on labels can be enough to make anyone's head spin. Rather than zooming in on what distinguishes one particular wine from the pack, it's far more useful to zoom out and learn about what all wines share in the big picture. Grasping a handful of central wine truths can help even a wine novice visualize how wines relate to one another in sensory terms. For instance, grapes pass through predictable stages of development as they ripen, and not all wines are made from grapes of equal ripeness. Degrees of ripeness correlate with many sensory factors—from wine's alcohol content, or weight, to oakiness and acidity. This kind of expert-level insight is the key not only to making educated guesses about how different wines will taste but also to predicting how they will pair with various foods.

VISUALIZING
WINE STYLE

THE RIPENESS FACTOR One advantage wine professionals have over other wine drinkers is that they can picture how wines relate to one another in terms of how they taste, not just by their ingredients or regional provenance. The inherent biology of grapes and yeast limits the range of possible winemaking outcomes, creating consistent patterns in flavor profile that anyone can learn to recognize. A few basic insights into factors like how grapes ripen can help beginners make educated guesses about which wines will be lighter or heavier, milder or bolder, sweeter or drier, before they open the bottle.

THE WINE STYLE SPECTRUM

Professionals know that there are clear and consistent patterns that govern wine style, and they use this knowledge to make educated guesses about how wines will taste before they open the bottle. You don't need to memorize dozens of grapes or regions to be able to use the same generalizations when you shop for wine.

MAPPING WINES BY STYLE

Navigating the wine world requires a sense of its range and boundaries. Wines can be loosely classified by their two most important power scales: weight and flavor intensity. Charting them on a classic grid can provide any wine drinker with expert-level insights into the big picture—into how wine styles relate to one another on a sensory level.

But why weight and flavor? Both are key sensory characteristics that are highly relevant to personal tastes and easy for beginners to identify. These traits correlate strongly with one another, as well as with other relevant wine qualities, such as acidity, oak, and tannin. Most important, though, they also have reasonably direct relationships with wine features that we can observe before we open the bottle, like color and alcohol content.

CRACKING THE WINE CODE

Charting wine styles by weight and flavor reveals some consistent patterns. Paler wines are often both lighter-bodied and milder in flavor than darker ones. White wine styles range more widely than reds. Pink rosé wines share territory with both whites and lighter red wines. Wines with bubbles are usually low in alcohol.

Such conclusions might seem banal, but just scratch the surface. Looking deeper into how wine styles relate to one another from a sensory perspective helps explain which factors have the greatest impact on relevant characteristics. These fundamentals help wine professionals navigate the wine world—and they can help you make more informed wine decisions, too.

CHARTING MAIN TRAITS

For the novice, it's very useful to picture where wines fall on a spectrum of style defined by wine's weight and flavor intensity. This provides a meaningful context for comparing wine options and for remembering how they taste.

WEIGHT

Heavier wines are:
Higher in alcohol • Richer in texture.
They are also often, but not always:
Bolder in flavor • Given oak treatment • Aged before bottling • Lower in acidity • From warmer regions • May be fortified with distilled spirit

Lighter wines are:
Lower in alcohol • More sheer in texture.
They are also often, but not always:
Milder in flavor • Unlikely to be given oak treatment • Bottled and sold young • High in acidity • From cooler regions • May be carbonated—sparkling or spritzy

FLAVOR

Milder wines are:
More subtle in flavor and scent, often with herbal/earthy flavors • Rarely given oak treatment.
They are also often, but not always:
Lower in alcohol • Paler in color • Bottled and sold young • High in acidity • From cooler regions

Bolder wines are:
More intense in flavor and scent, often with baked/spicy flavors • Often given oak treatment.
They are also often, but not always:
Higher in alcohol • Deeper in color • Aged before bottling • Lower in acidity • From warmer regions

RECOGNIZING PATTERNS IN HOW WINES TASTE AND RELATE TO ONE ANOTHER CAN HELP US MAKE INFORMED DECISIONS WITHOUT GETTING BOGGED DOWN IN EXCESSIVE DETAILS.

WHITE WINE
like Chardonnay
and Pinot Grigio

SPARKLING WINE
like Champagne
and Prosecco

ROSÉ WINE
like Anjou
and Tavel

RED WINE
like Shiraz
and Chianti

FORTIFIED WINE
like Port
and Sherry

HEAVIER

WEIGHT SCALE
A wine's weight is its perceived
thickness, or "body." It has a direct
relationship with alcohol content
in dry wines.

WEIGHT

LIGHTER

FLAVOR SCALE
The flavor intensity of a wine
is its concentration of aroma
compounds, which we
perceive as both flavor and
scent. It includes wine's fruit
and oak components.

MILDER   **FLAVOR**   BOLDER

THE THREE FLAVOR FACTORS

Wines are often labeled by grape variety, but ingredients alone can't tell you how a wine will taste. There are two additional factors that can play just as strong a role in shaping wine's flavor and style.

SAME GRAPE; DIFFERENT TASTE

In any wine's taste, the type of grape is just a starting point that will be shaped and changed by two more powerful forces: the vineyard's natural environment and the human vintner's active interventions. Consequently, wines made from the same grape variety can taste very different when grown in different regions and made using different methods. The three variables that control how any given wine will taste are:

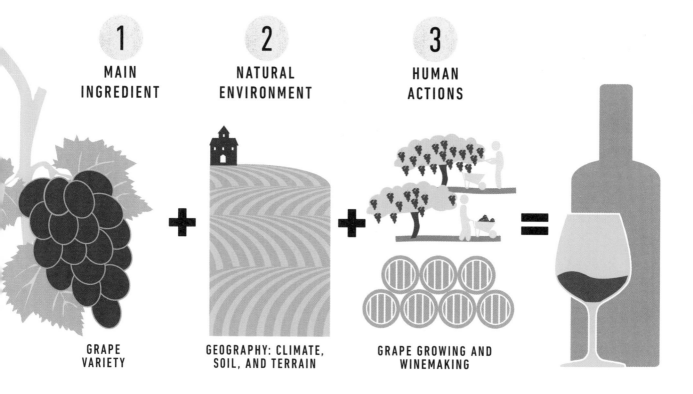

1

MAIN INGREDIENT

+

2

NATURAL ENVIRONMENT

+

3

HUMAN ACTIONS

=

GRAPE VARIETY

GEOGRAPHY: CLIMATE, SOIL, AND TERRAIN

GRAPE GROWING AND WINEMAKING

THE SAME GRAPE VARIETY CAN PRODUCE WINES THAT TASTE RADICALLY DIFFERENT, DEPENDING ON WHERE THEY'RE GROWN AND HOW THEY'RE MADE.

UNEXPECTED SIMILARITIES

We latch on to grape varieties when we shop for wine because they are prominent on wine labels, but also because they seem to offer a clear-cut way to decode wine. However, navigating solely by grape doesn't take into account the other two flavor factors. For instance, Pinot Noir and Syrah make vastly different wines. However, a French Pinot Noir from Burgundy and a French Syrah from the nearby Rhône will resemble each other more closely than they would

New World versions from California and Australia respectively. Common ground between the French wines in culture and climate will outweigh the family resemblance that is conveyed by the grape variety.

For similar reasons, a rich Chardonnay from California will be closer in flavor profile to Alsace Pinot Gris than it would to a lighter Chardonnay such as French Chablis, simply because both Alsace and California are sunny and warm. Chilly Chablis produces wines with more in common with other cool-climate whites, like Northern Italian Pinot Grigio.

ALL OVER THE MAP
Wines from different grapes grown in nearby regions or similar climates often taste more alike than wines from the same grape grown in vastly different regions or climates.

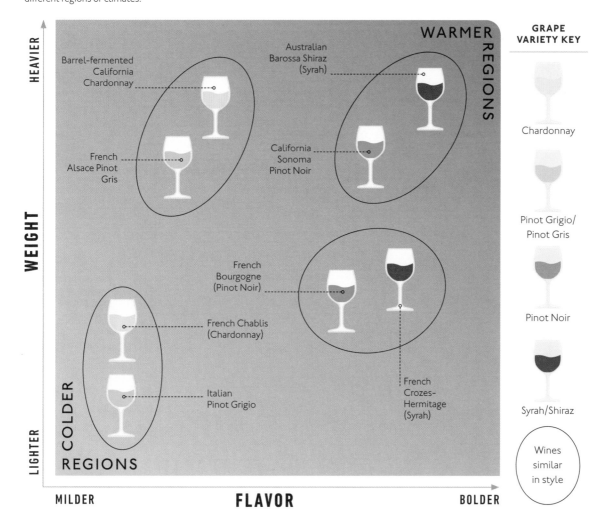

THE TASTING

Identifying the Style Spectrum

COMPARE EIGHT WINES AT HOME

Sample these four white wine styles and four red styles side by side, noting how they compare to one another in terms of body and overall flavor intensity. Consider gathering a group to learn together or freezing your remainders to enjoy later—see Freezing Wine on p.63. You can also taste whites and reds in separate sittings, though this makes it hard to discern relationships across colors.

1

VERY LIGHT,
BOLDEST FLAVOR

2

LIGHTWEIGHT,
MILD FLAVOR

3

MID-WEIGHT,
MEDIUM FLAVOR

4

HEAVYWEIGHT,
MEDIUM FLAVOR

MOSCATO D'ASTI

For example ... Italian, or Italian-style Moscato from Australia or California

About this wine This outlier on the spectrum is made from an uncommonly aromatic variety. Moscato's unusual pungency leads to flavorful wines even at low levels of alcohol

PROSECCO

For example ... Sparkling Italian styles from the Veneto or Venezie

About this wine A delicate, subtle-tasting, low-ripeness style— ideal for refreshing the senses in warm weather and daytime drinking.

SAUVIGNON BLANC

For example ... Styles from the New Zealand region of Marlborough

About this wine Noticeably richer in texture and more intense in aroma and flavor, thanks to greater ripeness and an aromatic grape variety.

BARREL-FERMENTED CHARDONNAY

For example ... Premium California styles from Sonoma County, Russian River Valley, Santa Barbara, or Carneros

About this wine Opulently rich in mouthfeel, thanks to ample sunshine, and flavor-boosted with the help of toasty new oak barrels.

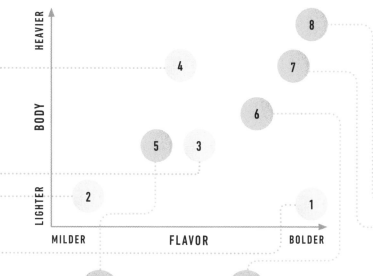

WINE LOVER'S TREASURE MAP

Here, some popular wines are plotted on our chart of weight and flavor intensity. Where do your favorites fall: bigger and bolder, at top right; or milder and more refreshing, at lower left? As you encounter wines, visualizing how they relate to others in sensory terms will help you look beyond grapes and regions in exploring your personal tastes.

5 MID-WEIGHT, MEDIUM FLAVOR

6 MID-WEIGHT, BOLD FLAVOR

7 HEAVYWEIGHT, BOLD FLAVOR

8 VERY HEAVY, BOLDEST FLAVOR

FRENCH PINOT NOIR

For example …
Bourgogne (Burgundy), or similar styles such as Mercurey or Santenay

About this wine This classic style is rather light and low in ripeness for a red, though mid-weight in the big picture, with restrained but seductive aromas.

SPANISH RIOJA

For example … Rioja Crianza, or similar but stronger Rioja Reserva

About this wine Riper and stronger, but not much heavier in weight. The flavor intensity is amplified by a strong presence of new oak scent.

PREMIUM SHIRAZ

For example …
Premium Australian Shiraz from the Barossa or McLaren Vale

About this wine
Extreme flavor density and body for a non-fortified wine, thanks to exceptional ripeness in sun-drenched South Australia.

FORTIFIED PORT

For example …
Portuguese Late-Bottled Vintage Port, or a Port-style fortified red California Zinfandel

About this wine This outlier is spiked with distilled grape brandy, making for a potent wine whose added alcohol turbo-charges its heft and concentration of flavor.

WHY REGIONS TRUMP GRAPES

Wine characteristics such as color, flavor, and alcohol content are almost direct reflections of the color, flavor, and sugar content of the grapes used. Geography and climate affect the development of these traits in the vineyard as the fruit grows and ripens on the vine.

THE SLIDING SCALE OF RIPENESS

As a general rule, most wine characteristics grow stronger together, regardless of the grape variety used to make them. A few others tend to decrease in similar progression. This pattern relates directly to the amount of sun grapes get in the vineyard and its effect on the ripening process.

WEIGHT INCREASES

ALONG WITH: FLAVOR INTENSITY • COLOR • COLOR DEPTH • JAMMY, "BAKED FRUIT" FLAVORS • EXOTIC, "SPICY" FLAVORS • OAK FLAVORS • TANNIN • ALCOHOL

SPARKLING WINE

WHITE WINE

ROSÉ WINE

RED WINE

FORTIFIED WINE

ACIDITY DECREASES

ALONG WITH: HERBAL, "GREEN FRUIT" FLAVORS • OUTDOORSY, "EARTHY" FLAVORS • CARBONATION

COMPARED TO GRAPE VARIETIES AND WINEMAKING, THE SUN HAS A MORE DRAMATIC IMPACT ON GRAPE RIPENESS AND, THEREFORE, ON A WINE'S FINAL STYLE.

FOR RED WINES AND HEAVIER WINES, GRAPES MUST BE VERY RIPE

Because their grapes need lots of sunshine and warmth, red wines and full-bodied wines tend to come from places that are very sunny, warm, and dry

To maximize ripeness and flavor potential, vintners usually let the fruit hang on the vine as long as possible.

FOR WHITE WINES AND LIGHTER WINES, GRAPES SHOULD NOT BE TOO RIPE

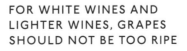

Since their grapes can suffer from too much sunshine and warmth, white wines and light-bodied wines more often hail from places that are cooler, cloudier, and more humid.

To avoid excessive ripeness and retain freshness, vintners often harvest the fruit earlier.

RIPENESS: A KEY CONCEPT

There is no single idea as powerful as ripeness for explaining how the wine world works and predicting how different wines will taste.

FROM TASTING "GREEN" TO TASTING GREAT

Ripeness is the final stage of fruit development, when it becomes ready to pick, with the right balance of flavors to taste fresh and delicious. Ripening shifts fruit from a hard, sour, immature stage toward a sweet, juicy state, accompanied by a color change from green to the fruit's proper color. We use the word "green" to describe the tastes associated with under-ripeness—sourness, bitterness, and the leafy flavor of vegetables—even though some fruits, like Granny Smith apples or white grapes, are still green in color when they are ripe and sweet. Plants get their energy from sunlight, through photosynthesis, so the degree of ripeness achieved by any fruit will depend on how much sun it gets in the final weeks before harvest.

THE FINAL PHASE OF ANY FRUIT'S GROWTH IS TO BECOME RIPE— SWEET AND READY TO EAT.

MORE HOURS OF SUNLIGHT AND WARMTH

MANY SHADES OF RIPENESS

For winemakers, picking grapes at exactly the right moment is critical, because it locks the flavor profile of their raw material in place. Sugar content is the main consideration in deciding when to harvest, since it determines wine's potential alcoholic strength. However, many other components are also evaluated, such as the fruit's levels of acidity, flavor compounds, and tannins.

Technically, there is no single universal definition of "perfect ripeness" among winemakers, since each grape component responds a little differently to changes in geography, weather, and farming techniques. Grapes with as little as 18 percent sugar would be considered fully ripe for Riesling grown in Germany's chilly Mosel, but not for California Cabernet Sauvignon, where anything under

24 percent sugar would be seen as underripe. Also, winemakers can and do harvest earlier or later depending on what styles they want to make, picking earlier for white or sparkling wine to retain refreshing acidity, or waiting longer for color and flavor to develop in the skins when making red wines.

Luckily, wine drinkers do not need to adopt the winemaker's nuanced view of ripeness. For the purpose of navigating wine by style, you'll be better served making a useful generalization: Think of the grapes used for lighter, cooler-climate wine styles like Riesling or Prosecco as being less ripe, and those used for heavier, warmer-climate styles like Cabernet Sauvignon or Port as being more ripe.

MOVING TARGETS

Many changes occur as grapes ripen. Their berries become bigger, softer, and juicier, tasting sweeter and less sour; their flavors shift from mild and vegetal to intense and fruity; and if they are of a purple variety, their skins deepen in color.

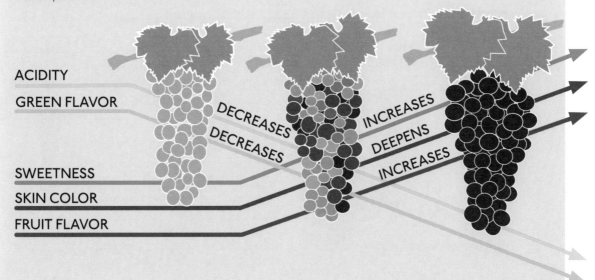

ACIDITY
GREEN FLAVOR
DECREASES
DECREASES
INCREASES
DEEPENS
INCREASES
SWEETNESS
SKIN COLOR
FRUIT FLAVOR

PREDICTING RIPENESS

It's often possible to deduce a wine's degree of ripeness from label clues like alcohol content, an essential step in cracking the code of the confusing wine world.

GIVE FLAVOR A BOOST

Grapes that receive more sun become riper, which increases the overall sensory impact of their wine on multiple levels. Sweeter grapes produce wines with higher alcohol when they are fermented all the way to dryness. Such wines feel heavier in the mouth but also amplify flavor by definition. Wines made from riper grapes tend to taste stronger because they have more flavor compounds such as aromatic esters but also because they contain more alcohol. Alcohol vaporizes easily, even at low temperatures, so a little extra alcohol acts as a scent and flavor booster in wine, just as it does in perfumes or liqueurs.

▲ **SECRET DECODER**
One of the most useful pieces of information on any wine label is hidden in the small print. Alcohol content correlates well enough to many wine traits to serve as a rough style indicator.

BELOW 13%

WINES TYPICALLY FEATURE LOW-RIPENESS TRAITS

Always lighter in texture
Usually higher in acidity
Often milder in flavor
Often paler in color
Rarely oaked
May be carbonated

EXAMPLES:
French Champagne
Spanish Albariño
Italian Chianti

BETWEEN 13% AND 14%

WINES TYPICALLY FEATURE MODERATE-RIPENESS TRAITS:

Mid-weight in texture
Moderate in acidity
Moderate in flavor
Moderate in color
May be oaked
Rarely carbonated

EXAMPLES:
Australian Chardonnay
French red Bordeaux
Oregon Pinot Noir

ABOVE 14%

WINES TYPICALLY FEATURE HIGH-RIPENESS TRAITS:

Always heavier in texture
Usually lower in acidity
Often bolder in flavor
Often deeper in color
Often oaked
Never carbonated

EXAMPLES:
California Zinfandel
Argentinian Malbec
French Châteauneuf-du-Pape

WHAT ALCOHOL CAN TELL YOU

In dry wines, where no grape sweetness is preserved, there is a nearly direct relationship between ripeness and alcohol content, and alcohol content must be listed on virtually all wine labels. We can predict a fair amount about how a given wine will taste just by knowing that 13.5% is the norm. We know that wines with higher alcohol will, by definition, be heavier in mouthfeel than average, but we can guess that they will also taste less tart and smell more intense and fruity due to greater ripeness. Dry wines with lower alcohol will usually be the reverse: lighter, milder, and more herbal.

The predictive power of alcohol content doesn't stop there. Some wine factors that are entirely under human control, such as degrees of oak and carbonation, are associated with higher or lower degrees of ripeness, and therefore alcohol content, for aesthetic reasons. The likelihood that a wine will be oaky increases greatly with higher-than-average alcohol, for example, while a lower-than-average alcohol level increases the chances of encountering carbonation. There are exceptions to these rules of thumb, of course, and wine qualities are least foreseeable in the crowded middle ground between 13% and 14%. But the patterns hold true enough to provide helpful guidance when wine shopping, and their predictive power becomes more ironclad the further alcohol levels deviate from the norm.

HOW GRAPE VARIETIES FIT IN

The type of grape used to make wine is a major style factor. Each has its own unique characteristics and flavor profile. Some grape varieties are very distinctive, while others are less easily recognizable. Like different varieties of apples or mangoes, wine grapes will look different and taste different when fresh. But since all vines depend on sunshine and ripeness for their fruit's development, grapes are not always wine's sole flavor factor.

ORGANIZING VARIETIES

In the same way that children's personalities become more apparent as they grow older, grape varieties often resemble one another at very low ripeness, but grow more and more distinct as they get riper. The traditional way to organize grape varieties is according to their native region—for example, Cabernet Sauvignon and Sauvignon Blanc both come from Bordeaux in France and are genetically related, while Chardonnay and Pinot Noir come from Burgundy. This certainly helps

WHITE GRAPE VARIETIES

Grapes with high flavor intensity are shown in **BOLD**.

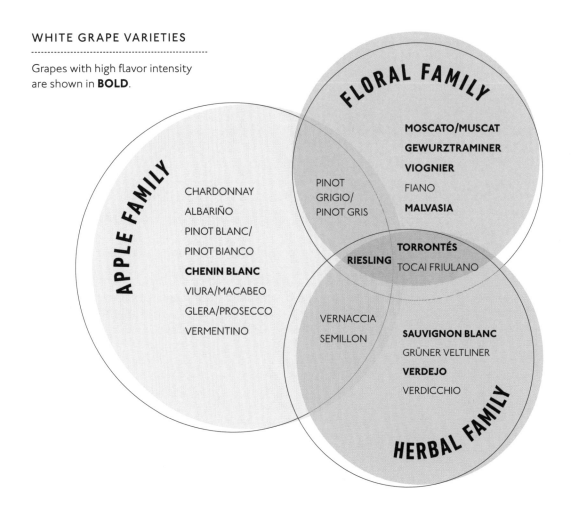

FLORAL FAMILY

APPLE FAMILY

CHARDONNAY

ALBARIÑO

PINOT BLANC/

PINOT BIANCO

CHENIN BLANC

VIURA/MACABEO

GLERA/PROSECCO

VERMENTINO

PINOT GRIGIO/ PINOT GRIS

MOSCATO/MUSCAT

GEWURZTRAMINER

VIOGNIER

FIANO

MALVASIA

TORRONTÉS

RIESLING   TOCAI FRIULANO

VERNACCIA

SEMILLON

SAUVIGNON BLANC

GRÜNER VELTLINER

VERDEJO

VERDICCHIO

HERBAL FAMILY

us navigate wine lists and retail stores, but from the wine drinker's perspective, it can be more helpful to classify grapes according to sensory qualities, particularly similarities in their overall flavor and scent.

OVERLAPPING FAMILIES OF FLAVOR

In white wines, there is a broad apple/pear resemblance between wines made from the most popular grape varieties; this is most obvious in the likes of Chardonnay and Pinot Grigio. But a few stand out with aromatics that are unusually distinct, like the leafy green scent of Sauvignon Blanc or the florality of Moscato. Some grapes combine elements of more than one family, such as Riesling, whose charming appley flavors seem to incorporate a touch of both flowers and herbs, like jasmine tea. In general, the stronger a white wine's aromatics, the less likely it is to be barrel-fermented and overtly oaked. For white wines, winemakers use new oak like a chef uses spices: to add personality to wines that have a subtle neutral scent.

MOST WHITE WINE GRAPES FEATURE AN APPLE-LIKE FLAVOR, BUT SOME PROJECT MORE INTENSE SCENTS, LIKE FLOWERS OR GREEN HERBS.

FLORAL FRAGRANCES

APPLE FLAVORS

LEAFY GREEN SCENTS

SORTING BY FLAVOR SIMILARITIES

Despite higher flavor intensity, red grapes can be harder to categorize aromatically than white grapes. Where white wines have simpler scents, reds are more complex, and most feature a layer of oak to some degree. However, red wine grapes can be sorted into a few broad "families."

Most red wines smell of fruits with deep colors, such as berries and cherries. Many of the most popular grapes smell and taste most like the darkest black fruits—say blackberry or blueberry—as with Cabernet Sauvignon and Malbec. A smaller number, like Pinot Noir and Sangiovese, taste more like brighter red berries, with scents reminiscent of strawberry or sour cherry. While most red wines tend to fall somewhere on the red fruit/ black fruit continuum, some feature an unusual concentration of additional scents—appetizing aromas and flavors that don't register as fruit, like black pepper and star anise. Since these grapes make wines that smell like they've been seasoned from the spice rack, such as Syrah and Grenache, we'll call this the spiced-fruit family.

LEFT TO STEW

Red wines get most of their flavor from steeping with dark grape skins during winemaking, a process that leads to stronger flavors and scents than those found in white wines.

RED GRAPE VARIETIES

Grapes with high flavor intensity are shown in **BOLD**.

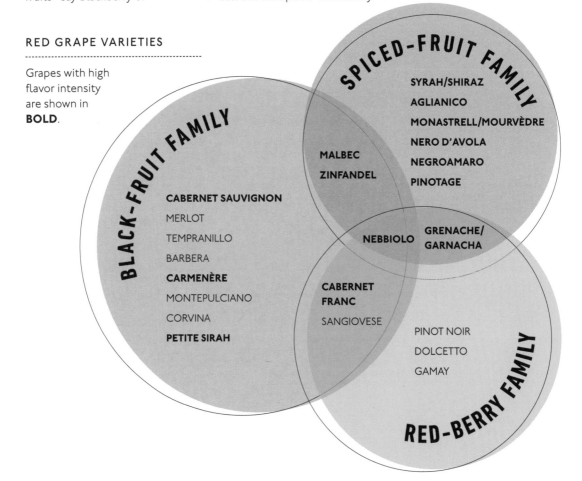

BLACK-FRUIT FAMILY

- **CABERNET SAUVIGNON**
- MERLOT
- TEMPRANILLO
- BARBERA
- **CARMENÈRE**
- MONTEPULCIANO
- CORVINA
- **PETITE SIRAH**

SPICED-FRUIT FAMILY

- **SYRAH/SHIRAZ**
- **AGLIANICO**
- **MONASTRELL/MOURVÈDRE**
- **NERO D'AVOLA**
- **NEGROAMARO**
- **PINOTAGE**

MALBEC
ZINFANDEL

NEBBIOLO **GRENACHE/ GARNACHA**

CABERNET FRANC
SANGIOVESE

RED-BERRY FAMILY

- PINOT NOIR
- DOLCETTO
- GAMAY

MANY POPULAR RED WINE
GRAPES PROJECT DENSE
BLACK-FRUIT FLAVORS,
BUT SOME FEATURE
BRIGHTER RED-BERRY
SCENTS OR A SPICED-FRUIT
AROMATIC PROFILE
MORE PROMINENTLY.

CHAPTER CHECKLIST

Here is a recap of some of the
most important points you've
learned in this chapter.

- Clear, consistent patterns govern **wine style**,
  and they can be used to make educated
  guesses about how a given wine will **taste**.

- A wine's taste is shaped by **three main
  factors:** the grape variety, the vineyard
  environment, and the decisions made by
  the winemaker.

- Wines from different grapes grown in **similar
  regions or climates** often taste more alike
  than wines from the same grape grown in very
  different regions or climates.

- **Ripeness** is the final stage of fruit
  development, when it becomes ready to pick
  and has the right balance of flavors to taste
  fresh and delicious.

- The ripening process shifts fruit from being
  hard and sour to a **sweet, juicy state**. There
  are also changes in color and flavor as grapes
  get more sun leading up to the harvest.

- Sunshine affects ripening fruit in predictable
  ways; some wine characteristics grow
  stronger together as grapes ripen, while a few
  others tend to **decrease** together.

- For red wines and **heavier wines**, grapes must
  be very ripe. For white wines and **lighter
  wines**, grapes should not be too ripe.

- Winemakers do not all share the same notion
  of **"perfect ripeness."** Each grape
  component responds differently to changes in
  geography, weather, and farming techniques.

- Riper grapes tend to make stronger-tasting wine
  because they have more **flavor compounds**,
  such as aromatic esters, but also because
  higher alcohol amplifies flavor.

- By knowing that **13.5% alcohol is the norm**,
  you can predict a fair amount about how a
  wine will taste based on its alcohol level.

**SPICED-FRUIT
FRAGRANCES**

**BLACK-FRUIT
FLAVORS**

**RED-BERRY
FLAVORS**

THE WHITE WINE SPECTRUM

EXPLORING BEYOND THE PALE White wines are incredibly diverse—they range from the most subtle Muscadet, to the most pungent Moscato; from the lightest, palest Riesling, to the heaviest Sherry, as dark as molasses. If we sort them by grapes and regions alone, as wine lists do, whites can seem like a disjointed and confusing category full of contradictions. But if we look at them from a different perspective, focusing on how they taste and relate to one another in sensory terms, clear patterns emerge. It becomes easier to explore and discover new wines to enjoy.

MAPPING WHITE WINES BY STYLE

Visualizing the relationships between wine styles is a helpful way to find and follow patterns in the tastes of any given subset of wines. Taken together, factors like grape and region, climate and winemaking approach, can give us a sense of which wines are likely to have the most in common in terms of sensory traits and flavor profile.

PLOTTING FOR WEIGHT AND FLAVOR

The graph below reveals helpful patterns among popular white wines, assessed by approximate weight and flavor. Notice that wines placed lower or to the left are from cool regions, while heavier and more intense wines shown farther up and to the right come from warmer places.

NEW WORLD WINES   OLD WORLD WINES

HEAVIER

SPANISH SHERRY

AMERICAN OAKED CHARDONNAY

FRENCH SAUTERNES

CALIFORNIA VIOGNIER

FRENCH MEURSAULT

FRENCH ALSACE GEWURZTRAMINER

AUSTRALIAN UNWOODED CHARDONNAY

ARGENTINIAN TORRONTÉS

SPANISH VERDEJO

AUSTRALIAN RIESLING

NEW ZEALAND SAUVIGNON BLANC

FRENCH MÂCON-VILLAGES

FRENCH CHABLIS

FRENCH WHITE BORDEAUX

FRENCH SANCERRE

FRENCH MUSCADET

AUSTRIAN GRÜNER VELTLINER

ITALIAN PINOT GRIGIO

SPANISH ALBARIÑO

CALIFORNIA MOSCATO

FRENCH VOUVRAY

FRENCH CHAMPAGNE

ITALIAN PROSECCO

PORTUGUESE VINHO VERDE

GERMAN RIESLING

LIGHTER

ITALIAN MOSCATO D'ASTI

MILDER   **FLAVOR**   BOLDER

WEIGHT

RAMPING UP FLAVORS

Some grapes naturally taste more intense and distinctive than others, but sunshine and warmth develop flavors in all grapes as they ripen. Generally speaking, the lower a wine's alcohol content, the milder and more neutral its flavor is likely to be. With the exception of a few hyper-aromatic grapes, this is particularly true of whites, which can taste quite pleasant at very low degrees of ripeness.

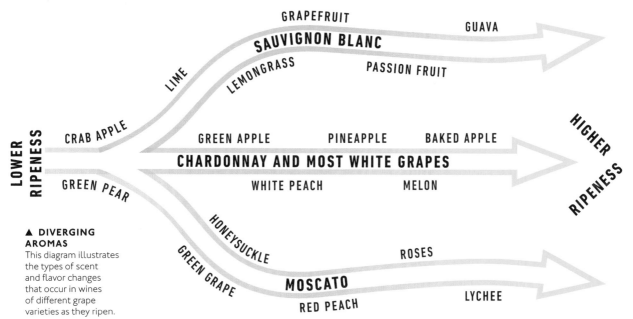

▲ DIVERGING AROMAS
This diagram illustrates the types of scent and flavor changes that occur in wines of different grape varieties as they ripen.

UNDERRIPE GRAPES ALL SHARE SIMILAR FLAVORS. SUN AND RIPENESS BRING OUT THE UNIQUE CHARACTER OF INDIVIDUAL VARIETIES.

BECOMING INDIVIDUALS

In colder regions or when harvested early, wines made from most white grapes share a similar low-ripeness aromatic profile, featuring mild, simple apple and pear flavors. Such wines also share other low-ripeness features—like low alcohol, high acidity, and absence of new oak flavors. As a result, these qualities almost always go hand in hand in the lower-left corner of the grid opposite.

When grapes achieve greater ripeness, either through warmer climate conditions or by delaying harvest, the individual aromatic personality of each variety becomes more apparent. Most white grapes get peachier and more tropical in flavor as they ripen, moving toward the top-right corner of the grid. Some may eventually reach a stage where their fruit tastes cooked or dried. Those varieties with the most distinctive aromatic character, like floral Moscato or herbal Sauvignon Blanc, get more intense with ripeness, too, but will progress in their own direction.

CHARDONNAY'S STYLE RANGE

Chardonnay, the most popular white wine grape on earth, is a great illustration of how wines made from the same variety can fall in different zones on the style chart. The primary controlling factor is the degree of ripeness, so geography and climate play a major role. However, winemaking decisions can further diversify the results, especially when flavor-boosting techniques are used.

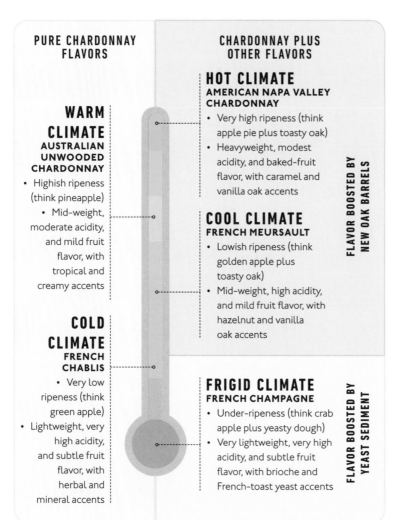

PURE CHARDONNAY FLAVORS

CHARDONNAY PLUS OTHER FLAVORS

WARM CLIMATE
AUSTRALIAN UNWOODED CHARDONNAY
- Highish ripeness (think pineapple)
- Mid-weight, moderate acidity, and mild fruit flavor, with tropical and creamy accents

COLD CLIMATE
FRENCH CHABLIS
- Very low ripeness (think green apple)
- Lightweight, very high acidity, and subtle fruit flavor, with herbal and mineral accents

HOT CLIMATE
AMERICAN NAPA VALLEY CHARDONNAY
- Very high ripeness (think apple pie plus toasty oak)
- Heavyweight, modest acidity, and baked-fruit flavor, with caramel and vanilla oak accents

COOL CLIMATE
FRENCH MEURSAULT
- Lowish ripeness (think golden apple plus toasty oak)
- Mid-weight, high acidity, and mild fruit flavor, with hazelnut and vanilla oak accents

FLAVOR BOOSTED BY NEW OAK BARRELS

FRIGID CLIMATE
FRENCH CHAMPAGNE
- Under-ripeness (think crab apple plus yeasty dough)
- Very lightweight, very high acidity, and subtle fruit flavor, with brioche and French-toast yeast accents

FLAVOR BOOSTED BY YEAST SEDIMENT

STRENGTH AND WEAKNESS

Like many white wine grapes, Chardonnay is understated in flavor and scent, delivering familiar appley flavors that vary based on ripeness. Unlike most other grapes, Chardonnay can provide wines of seductive texture with balanced acidity at radically different degrees of ripeness, making it capable of producing world-class wine in both the coldest and warmest wine regions. Its greatest weakness is a relatively neutral flavor profile, so winemakers often choose to add flavor to bring its aromatic intensity in line with its tactile richness. For still wines, barrel fermentation and aging add the toasty, dessert-spiced flavor of oak. For sparkling wines, a similar aging strategy is used, but with the winemaking yeast sediments, or lees, providing a bready, baked-goods flavor.

THE TASTING

Identifying Chardonnay's Stylistic Range

COMPARE THREE CHARDONNAYS AT HOME

Sample these wines side by side, paying special attention to which characteristics change as you move from colder to warmer wine regions and, therefore, from lower to higher degrees of grape ripeness within the same variety.

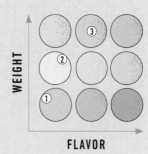

FLAVOR LOCATOR CHART

These Chardonnays are plotted on the chart according to their weight and flavor, which vary based on factors such as grape ripeness, winemaking practices, and cultural traditions..

WEIGHT

FLAVOR

1

LOW RIPENESS, NO OAK

2

MEDIUM RIPENESS, NO OAK

3

HIGH RIPENESS, NEW OAK

FRENCH WHITE BURGUNDY

For example ... Chablis or other unoaked white Burgundy, such as Mâcon-Villages, Rully, or Beaujolais Blanc

Can you detect ...?
Pale color;
very low sugar/very dry;
high acidity/tart;
low fruit intensity;
no oak flavor;
light to medium alcohol

NEW WORLD UNOAKED CHARDONNAY

For example ... Australian unwooded or an unoaked New World option from California, Chile, or South Africa.

Can you detect ...?
Pale color;
low sugar/dry;
medium acidity/tangy;
low fruit intensity;
no oak flavor;
medium to high alcohol

NEW WORLD OAKED CHARDONNAY

For example ... Napa Valley Chardonnay or other barrel-fermented New World example over 14% alcohol from California, Washington, Chile, South Africa, or Australia

Can you detect ...?
Golden color;
low sugar/dry;
medium acidity/tangy;
medium fruit intensity;
overt oak flavor;
high alcohol

EXPLORING LIGHTER WHITE STYLES

Truly light-bodied wines are those with 12.5% alcohol or less—wines that are almost invariably white and consistently share a sheer, delicate mouthfeel. When dry, lightweight wines are made from low-ripeness fruit by definition, but many are lightly sweet wines whose fermentation was stopped to retain some sugar. However, fully sweet dessert wines are an exception: Many are heavier in texture than their alcohol alone would suggest.

IF YOU ENJOY
THESE LIGHTER
WHITE WINES ...

SHEER DELIGHTS

The lightest white wines typically feature high degrees of refreshing acidity, and the category includes almost all sparkling and semi-sparkling wines.

Those that are driest tend to be mildest in flavor, because they are made with fruit of low ripeness. To acquire bolder flavors, lightweight wines must be made either from an aromatic grape variety, such as Moscato or Riesling, or in a lightly sweet style, where greater ripeness is achieved but not all grape sugar gets converted to alcohol—or sometimes both at once.

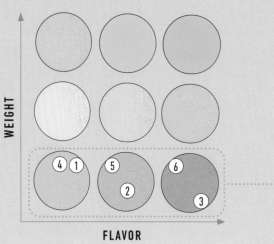

... TRY THESE
WINES WITH
SIMILAR SENSORY
PROFILES.

MILDER FLAVOR

MEDIUM FLAVOR

BOLDER FLAVOR

1 FRENCH BRUT CHAMPAGNE

Sparkling wines are a specialty of cool regions, like Champagne in northern France. Their refreshing fizz and acidity, as well as their subtlety of flavor, are hallmarks of grapes of low ripeness, often a mix of both green- and red-skinned varieties.

2 GERMAN MOSEL RIESLING

Mosel Rieslings are the lightest of the world's truly noble wines. Most feature very low alcohol and a sweet/tart flavor profile. Riesling is an aromatic variety whose wines taste stronger as its grapes grow riper and sweeter.

3 ITALIAN MOSCATO D'ASTI

Moscato grapes are freaks of nature that deliver an uncanny perfume-like intensity of flavor at even modest degrees of ripeness. Asti styles are only partially fermented, so their grapes ripen further than alcohol alone might suggest.

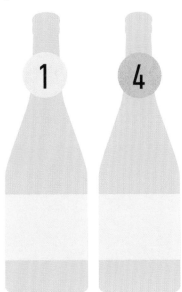

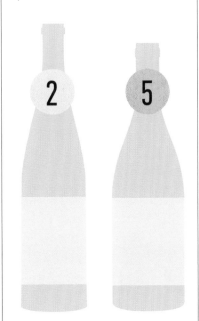

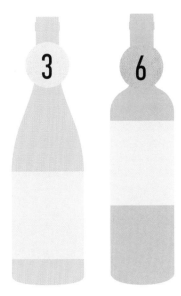

4 SPANISH CAVA

Catalan sparkling wines are made with native Spanish grapes but follow the methods pioneered in France for making Champagne. The resulting wines often feature less finesse but can deliver their own unique rich appeal and have more tempting price tags.

5 FRENCH VOUVRAY

The Chenin Blanc grape of Vouvray shares many of Riesling's distinctive traits. Grown in the chilly Loire Valley, it makes wines that are not as delicate as the Mosel's but are similar in sweet/tart balance and appley aromatic intensity.

6 PORTUGUESE VINHO VERDE ROSADO

No other wine has Moscato's flavor profile, but its fans often enjoy these pink fizzy refreshers from northern Portugal. Milder in flavor and less overtly sweet, they are made from red grapes harvested while still "green."

EXPLORING MID-WEIGHT WHITE STYLES

Given the right growing conditions, any white grape can produce a mid-weight wine, and at least three out of every four white wines will fall into this category, over 12.5% alcohol but below 14%. This represents the natural range of dry wines from grapes of normal ripeness and includes styles that are very popular, thanks to their emphasis on refreshment and food-flattering qualities.

IF YOU ENJOY THESE MID-WEIGHT WHITE WINES ...

HAPPY MEDIUM

Being the most crowded of white wine categories, this is also the most diverse and the one where the grape variety plays the strongest role.

Mid-weight wines are almost always dry and range from moderately tangy to very tart in acidity. Fruit flavor intensity in this category depends largely on the grape's aromatic power, but some wines are flavor-boosted during winemaking with oak or aging. A mild touch of new oak flavor from barrel fermentation is common for mid-weight Chardonnays but less so for wines made with other grapes.

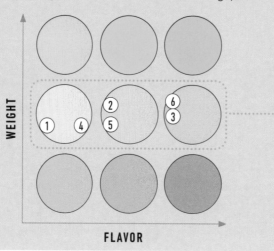

... TRY THESE WINES WITH SIMILAR SENSORY PROFILES.

MILDER FLAVOR

MEDIUM FLAVOR

BOLDER FLAVOR

1 NORTHERN ITALIAN PINOT GRIGIO

White wines are prized more for refreshment than richness in Italy, so this style is harvested early. The wines feature modest alcohol and snappy acidity as a result, as well as mild, understated flavors.

2 FRENCH BOURGOGNE BLANC

Many popular white Burgundies feature a light kiss of toasty oak to supplement Chardonnay's subtle flavor, but not all. They are lighter and more understated than New World Chardonnays and are known for their finesse.

3 NEW ZEALAND SAUVIGNON BLANC

Sauvignon Blanc is grown throughout the New World but most notably in this island nation. Its cool climate is perfectly suited for developing intense citrus and herbal flavors in this aromatic grape variety.

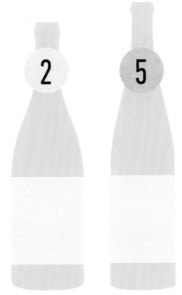

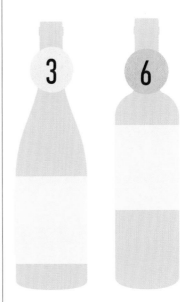

4 SPANISH ALBARIÑO

This sophisticated white comes from Galicia on Spain's Atlantic coast and is tailor-made for seafood. It is similar to Pinot Grigio in body and subtlety, but it has sharper acidity and more fragrant aromatics.

5 AUSTRIAN GRÜNER VELTLINER

This Austrian grape is known for its "green" scents and vibrant acidity. Aromas of celery and white pepper distinguish the wines, which range from sappy, fruity refreshers to deluxe wines of substance.

6 ARGENTINIAN TORRONTÉS

Rarely seen outside Argentina, the Torrontés grape makes wines that share the intensity of flavor and dryness associated with Sauvignon Blanc but smell more like floral Moscato.

EXPLORING HEAVIER WHITE STYLES

There are far fewer white wines than reds in the heavyweight category, at 14% alcohol or higher, because to reach this level they must either be made from hyper-ripe grapes or be strengthened with distilled spirit. Chardonnays dominate the dry range of full-bodied whites, but this segment also includes most sweet dessert wines and all fortified whites, which can contain up to 20% alcohol.

IF YOU ENJOY THESE HEAVIER WHITE WINES ...

HEAVY HITTERS

White wines in this category always feature dense, rich texture but can range all over the map in sweetness—from completely dry to syrupy sweet.

Acidity is rarely high outside the finest wines. The near-direct correlation between alcohol content and fruit flavor in wine naturally skews heavyweight wines toward the bolder end of the aromatic spectrum. Flavor intensity may be inherent to the grape in some cases but is more often amplified in the winery or vineyard—by fermenting in new oak barrels, for example, or using super-ripe, late-harvest fruit.

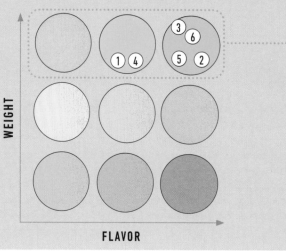

... TRY THESE WINES WITH SIMILAR SENSORY PROFILES.

 **MILDER FLAVOR**

 **MEDIUM FLAVOR**

 BOLDER FLAVOR

1 AUSTRALIAN OAKED CHARDONNAY

Chardonnays from sunny New World regions are often oaked—through barrel fermentation for premium wines, and oak chips for modest wines. This is warranted by the fuller body and greater ripeness routinely achieved in places like Australia.

2 FRENCH ALSACE GEWURZTRAMINER

The sunny Alsace region's most aromatic wine is the peach-and-lychee–scented Gewurztraminer. Here, this grape makes lush, power-house wines that taste surprisingly dry with overtly floral flavors and uncommonly low levels of acidity.

3 SPANISH MEDIUM SHERRY

The world's strongest white wines are fortified Sherries from Andalusia in Spain. They are spiked with brandy, then flavor-boosted with yeast and oxidation in a unique aging process. Some are also sweetened with raisin syrup to complement their nutty taste.

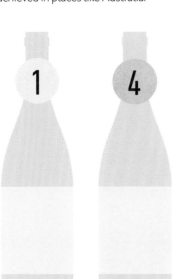

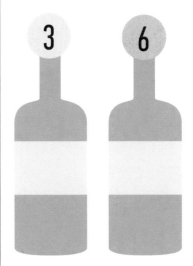

4 SOUTH AFRICAN OAKED CHENIN BLANC

Few wines can deliver the one-two punch of alcohol and oak of Chardonnay without losing acidity and balance. Chenin Blanc can do it with grace in warm, dry regions, such as South Africa's coastal wine lands, and some are lightly sweet.

5 CALIFORNIA VIOGNIER

Viognier is a French grape from the Rhône region but is now more often grown in New World regions like California. Like a mashup of Chardonnay and Gewurztraminer, it makes gorgeously luscious wines with effusive floral aromas.

6 PORTUGUESE RAINWATER MADEIRA

The tropical island of Madeira makes a unique fortified white wine whose Sherry-like flavors are intensified and caramelized by applying heat during aging. This uncommon technique gives the wine a sharply nutty taste.

THE RED
WINE
SPECTRUM

EXPLORING THE DARK SIDE There are just as many distinct and delicious red wine styles as there are white. However, the logistics of making red wines look red lead this category of wines to have a narrower range in terms of alcohol content and flavor intensity than white wines do. Winemakers blend multiple grape varieties together more often in red wines than whites, making it even harder to decide what to drink. Luckily, a few key concepts can help wine drinkers anticipate what qualities they'll find in the glass—whether it's pure, pale Pinot Noir or inky, blended Cabernet Sauvignon; cool-climate Chianti or sun-loving Shiraz.

MAPPING RED WINES BY STYLE

There is no doubt that red wines have lots of personality and individual character, but they have less stylistic diversity than whites in terms of body and flavor intensity. A handful of whites can be as bold or heavy as the heftiest reds, but no red wines are as subtle or delicate as the lightest, mildest whites. Therefore, in discussing red wines, we'll focus on the upper right-hand quadrant of our style-plotting grid—red wine's bolder-flavored, richer-textured territory.

SKINS LIMIT RED WINE RANGE

The purple peels of dark grapes give red wines their color, as well as the flavors that make red wines so aromatically diverse. However, the growing conditions needed to develop these compounds and the logistics of transferring them from the grape skins to the wine rule out the possibility of making as broad a range of red wine styles as we see in white.

- **Flavor Intensity** The same compounds in dark grape skins that supply color in red wines also supply flavor. Since both come from the same source and can't be separated, wines that look red cannot taste as mild as whites can.

- **Weight** Purple grapes need more sun to ripen than green grapes. At the low degrees of ripeness that make truly light-bodied wines, the peels of dark grapes are low in color and fruit flavor and high in bitterness and vegetal flavors. As a result, there is little commercial demand for red wines below 12.5% alcohol.

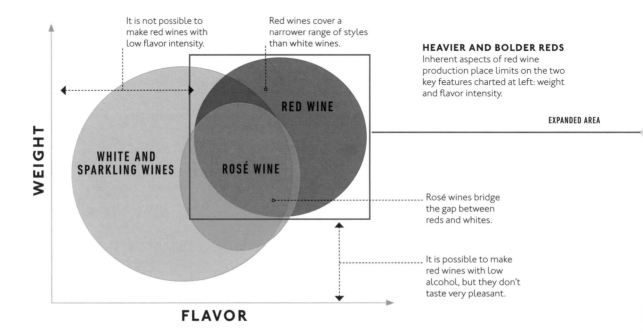

It is not possible to make red wines with low flavor intensity.

Red wines cover a narrower range of styles than white wines.

HEAVIER AND BOLDER REDS
Inherent aspects of red wine production place limits on the two key features charted at left: weight and flavor intensity.

EXPANDED AREA

WEIGHT

WHITE AND SPARKLING WINES

RED WINE

ROSÉ WINE

Rosé wines bridge the gap between reds and whites.

It is possible to make red wines with low alcohol, but they don't taste very pleasant.

FLAVOR

PLOTTING FOR WEIGHT AND FLAVOR

On the graph below, the relationships between key red wine styles are illustrated, with circles approximating their stylistic range. There are always exceptions, of course, with top wines often being heavier and more intense than the averages shown.

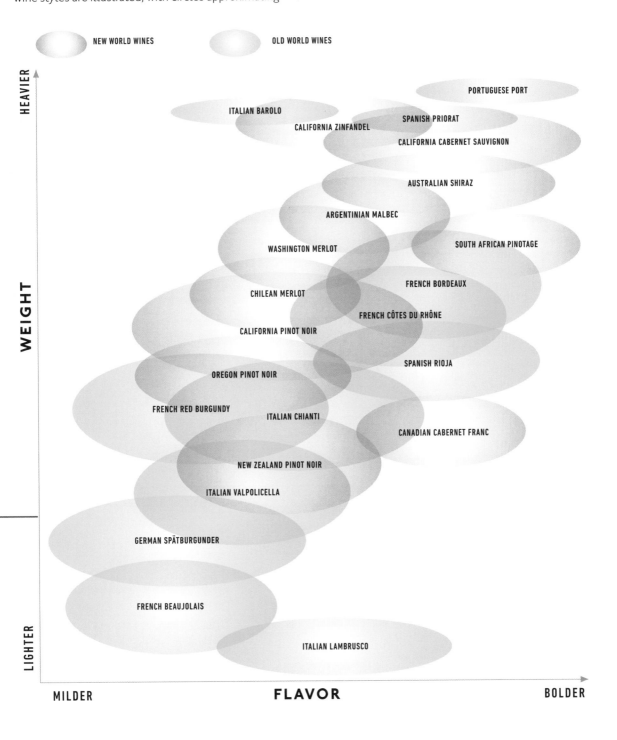

NEW WORLD WINES   OLD WORLD WINES

HEAVIER

WEIGHT

LIGHTER

PORTUGUESE PORT

ITALIAN BAROLO

CALIFORNIA ZINFANDEL

SPANISH PRIORAT

CALIFORNIA CABERNET SAUVIGNON

AUSTRALIAN SHIRAZ

ARGENTINIAN MALBEC

WASHINGTON MERLOT

SOUTH AFRICAN PINOTAGE

CHILEAN MERLOT

FRENCH BORDEAUX

FRENCH CÔTES DU RHÔNE

CALIFORNIA PINOT NOIR

SPANISH RIOJA

OREGON PINOT NOIR

FRENCH RED BURGUNDY

ITALIAN CHIANTI

CANADIAN CABERNET FRANC

NEW ZEALAND PINOT NOIR

ITALIAN VALPOLICELLA

GERMAN SPÄTBURGUNDER

FRENCH BEAUJOLAIS

ITALIAN LAMBRUSCO

MILDER   **FLAVOR**   BOLDER

RED WINE FLAVOR PROGRESSION

Red wines have more in common with one another in big picture traits like body and flavor intensity than whites. However, they tend to be more aromatically distinct because most grape flavor compounds are found in the peel, and only red wines are fermented with their skins (see pp.138–139). Much of the variation derives from the particular grape variety used, but the familiar pattern of ripeness also comes into play when considering red wines as a group.

MAKING EDUCATED GUESSES

The alcohol content of red wines is a better proxy for grape ripeness and associated characteristics than it is in whites because reds are more reliably fermented dry. The standard pattern—low-ripeness wines smelling more subtly of sour fruits and green herbs, and high-ripeness wines smelling more intensely of cooked fruit and dessert spices—is also more apparent because red wines taste stronger overall, and being served warmer amplifies their aromatics. The color saturation of red wines is an indicator of ripeness as well, reducing the need to rely on alcohol alone.

COMMON RED WINE SCENTS AND FLAVORS

Many aspects of a wine go hand in hand. This diagram shows which general categories of scents and flavors are associated with different levels of ripeness in red wines, as well as how these correlate to other style factors.

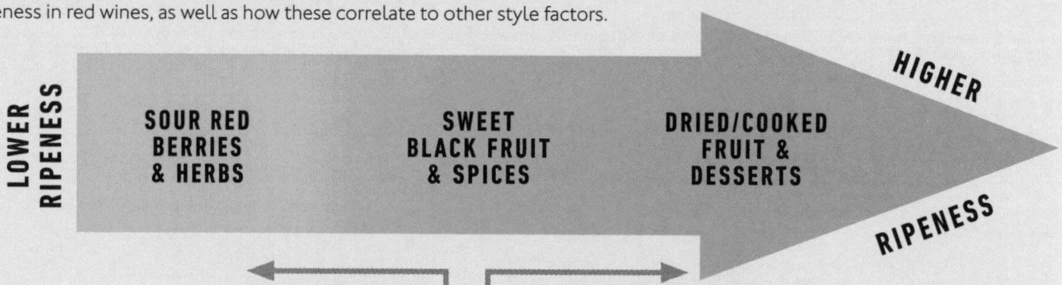

LOWER RIPENESS

HIGHER RIPENESS

SOUR RED BERRIES & HERBS

SWEET BLACK FRUIT & SPICES

DRIED/COOKED FRUIT & DESSERTS

ASSOCIATED WITH:
Alcohol below 13.5%
Colder wine regions
Traditional European wines

OFTEN OCCUR ALONGSIDE:
High levels of acidity
Pronounced dryness
Low levels of new oak flavor
Harsh "green" tannins (think astringent tea)

ASSOCIATED WITH:
Alcohol over 14%
Warmer wine regions
Modern New World wines

OFTEN OCCUR ALONGSIDE:
Lower levels of acidity
Moderate dryness
High levels of new oak flavor
Velvety "soft" tannins (think plush hot cocoa)

KEEPING IT IN THE FAMILY

The most obvious common ground in red wine flavors is found in closely related grapes, such as the three main Bordeaux varieties. Recent genetic studies show a parent/offspring relationship between Merlot and Cabernet Franc, and that Cabernet Sauvignon's parents are Cabernet Franc and Sauvignon Blanc. Wines from these grapes share a leafy, vegetal flavor profile in cold regions and wet vintages but become easier to tell apart with more sunshine and ripeness.

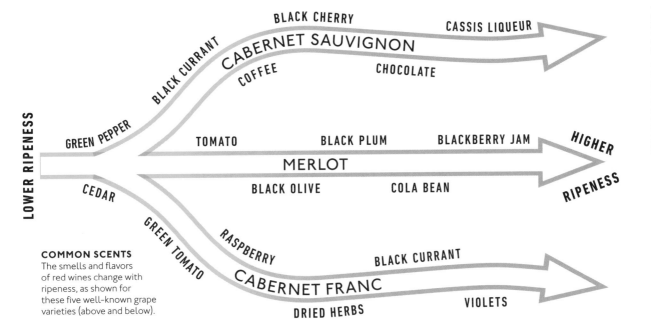

COMMON SCENTS
The smells and flavors of red wines change with ripeness, as shown for these five well-known grape varieties (above and below).

UNEVEN RIPENING

Not all grapes ripen alike. Even at their ripest, cool-climate varieties like Pinot Noir rarely achieve the same levels of flavor development routinely reached in warm-climate grapes like Shiraz.

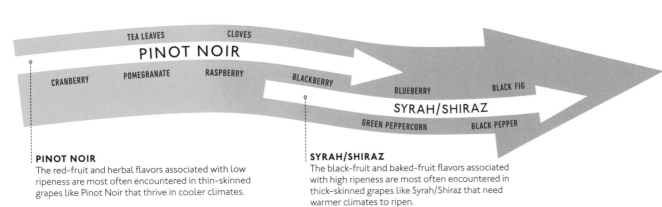

PINOT NOIR
The red-fruit and herbal flavors associated with low ripeness are most often encountered in thin-skinned grapes like Pinot Noir that thrive in cooler climates.

SYRAH/SHIRAZ
The black-fruit and baked-fruit flavors associated with high ripeness are most often encountered in thick-skinned grapes like Syrah/Shiraz that need warmer climates to ripen.

CABERNET SAUVIGNON'S STYLE RANGE

Wines made with Cabernet Sauvignon provide a great example of how and why winemakers often combine multiple grapes. Since this variety makes darker, more flavorful wines than most others, it is often used in blends. Cabernet Sauvignon reliably boosts wine's weight, flavor intensity, color depth, and ageability, allowing vintners more creative control.

BORDEAUX BLENDS

Blending grapes is traditional in the Bordeaux region, where native Cabernet Sauvignon is most respected but its milder cousin Merlot is more widely planted. Cabernet Sauvignon struggles to ripen in cooler zones, making wine that is dark and strong but tannic. In sites with more ripening potential, such as Bordeaux's Left Bank, it can thrive and make wines that are even darker and stronger but taste more fruity and less harsh. Over centuries, vintners have adapted to this environment—using small amounts of Cabernet Sauvignon to intensify lighter Merlot-based wines in cooler sites and modest wines, and small amounts of Merlot to mellow the brute force of Cabernet Sauvignon in warmer sites and premium wines.

CABERNET SAUVIGNON CONTENT IN BLENDED WINES

Most wines labeled Cabernet Sauvignon are not 100 percent Cabernet Sauvignon, because most wine-labeling laws require only 75–85 percent of the named grape to be present. Only in the warmest, sunniest regions does this variety ripen thoroughly enough to be pleasing unblended.

<25%

50% TO 75%

75% TO 99%

100%

Traditional Bordeaux blend—modest A small proportion of Cabernet Sauvignon adds color, body, and flavor without dominating. Typically it boosts lighter wines in cooler regions, as in Merlot-based Bordeaux but also many Tuscan and Spanish reds.

Traditional Bordeaux blend—premium Cabernet Sauvignon is the primary ingredient but is partnered with grapes that can soften and flesh out its austere qualities. Many of the world's finest reds follow this model but cannot be labeled as Cabernet Sauvignon.

New World Cabernet Sauvignon blend When this grape ripens more fully, it becomes more intense and less harsh, as is common in the Americas and the southern hemisphere. Some blending is still the norm, but to a lesser degree and not always mentioned on labels.

Pure unblended Cabernet Sauvignon Cabernet must reach a high degree of ripeness to taste balanced on its own, especially in a premium wine. The combination of its natural intensity and high ripeness makes for some of the world's most concentrated wines.

THE TASTING

Identifying Cabernet Sauvignon Traits

COMPARE THREE BLENDED RED WINES AT HOME

Sample these wines side by side, paying special attention to how characteristics change between them. While percentages of Cabernet Sauvignon will vary, they are likely to be higher in heavier wines from warmer regions, where grapes ripen more fully.

BLENDING TO CONTROL WEIGHT

In cool regions like its native Bordeaux, small amounts of Cabernet Sauvignon are typically used to strengthen a mid-weight wine. In sunnier regions, like Chile or California, the roles are often reversed.

UP TO 75% CABERNET SAUVIGNON

UP TO 75% CABERNET SAUVIGNON

MORE THAN 75% CABERNET SAUVIGNON

FRENCH LEFT BANK BORDEAUX	CHILEAN BORDEAUX-STYLE BLEND	AMERICAN SONOMA CABERNET SAUVIGNON
For example ... Château wines from modest appellations like Graves, Haut-Médoc, Listrac, or Moulis	**For example ...** Blends where a grape variety is not named on the front label but where Cabernet Sauvignon is mentioned on the back label	**For example ...** A premium Sonoma appellation, like Alexander Valley or Knights Valley, or other California sources, like Napa Valley
Can you detect ...? Medium color; very low sugar/very dry; high acidity/tart; medium fruit intensity; mild oak flavor; medium alcohol/mid-weight; leathery tannic mouthfeel	**Can you detect ...?** Dark color; low sugar/dry; medium acidity/tangy; high fruit intensity; mild oak flavor; high alcohol/heavy; softer velvety mouthfeel	**Can you detect ...?** Very dark color; low sugar/dry; medium acidity/tangy; very high fruit intensity; strong oak flavor; high alcohol/heavy; softer velvety mouthfeel

EXPLORING LIGHTER RED STYLES

Since very few red wines are truly light-bodied, those that we think of as "lighter" in style tend to be those below 13.5% alcohol—a touch stronger than the range for lightweight whites. Outside some niche wines for entry-level audiences, vintners overwhelmingly ferment their red wines to dryness. This results in a fairly consistent relationship between alcohol content and ripeness among reds and fewer sweet exceptions.

IF YOU ENJOY THESE LIGHTER RED WINES ...

ON THE BRIGHT SIDE

Almost all lightweight reds are from cool climates, where lower-than-average grape ripeness is the norm.

They generally feature elevated acidity and bright flavors that lean toward the red-berry and herbal end of the aromatic spectrum. Many are young, value-oriented wines that spend little or no time in barrels, to preserve fruity freshness. But premium wines are likely to be oak-aged, just as for heavier reds. When vintners want to retain light sweetness or obtain lower alcohol, they usually opt to make rosé instead of red.

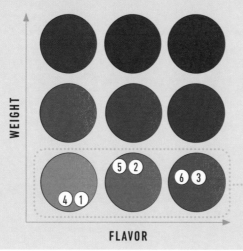

WEIGHT

FLAVOR

... TRY THESE WINES WITH SIMILAR SENSORY PROFILES.

MILDER FLAVOR

MEDIUM FLAVOR

BOLDER FLAVOR

1 FRENCH BEAUJOLAIS-VILLAGES

The Beaujolais district of Burgundy is known for its very light red wines made with Gamay grapes. Gamay has uncommonly low tannin, so its wines are among the few reds that taste pleasant chilled.

2 ITALIAN CHIANTI

This popular Italian red is made primarily with Sangiovese grapes, known for their high levels of acidity and tannin. While everyday Chianti may see little or no aging, the finest are stronger, more intense, and often aged longer.

3 AMERICAN OREGON PINOT NOIR

Pinot Noir makes wines that are lighter and paler than the average red but are some of the world's most sought after. Known for its seductive aromatics and silky texture, red Burgundy's noble grape thrives in cool regions like coastal Oregon.

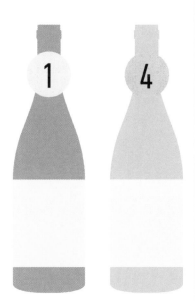

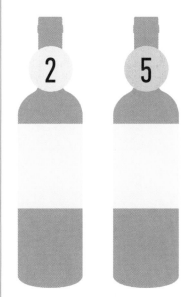

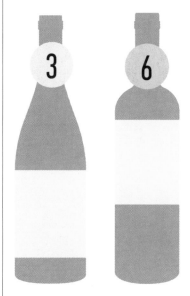

4 FRENCH TAVEL ROSÉ

To find wines as slurpable as Beaujolais, it's best to look to rosé wines like those made in the south of France. These medium-bodied Grenache-based blends are dry, with a flavor profile that is closer to red wine than white.

5 PORTUGUESE DOURO

Few regions produce red wines with enough acidity and tannin to rival Italian reds, but northern Portugal is one to watch. Douro wines are dry reds made in the same district as Port and from the same mix of native grapes.

6 SPANISH RIOJA

Spain's Tempranillo grape often makes denser, darker wines, but in cool Rioja its wines have a brighter, Pinot Noir–like appeal. The region's tradition of barrel aging results in some of the oakiest of lighter red wines.

EXPLORING MID-WEIGHT RED STYLES

The middle ground in red wines encompasses the majority of reds and skews a little stronger than in whites, with most containing between 13.5% and 14.5% alcohol. The modern consumer's preference for rich, flavorful reds leads vintners to shoot for this weight range, which has become the new normal only in recent decades. A century ago, the average red wine was considerably lighter than it is today.

IF YOU ENJOY
THESE MID-WEIGHT
RED WINES ...

WORLD OF DIFFERENCE

Mid-weight reds are the most diverse because so many grapes and appellations contribute to this popular category.

Most are dry wines, but they can be all over the map in aromatic character, color depth, and degree of oak flavor. A fairly consistent pattern of distinctions can be found, however, between most Old World styles and those from the New World. In general, the driest wines that are most austere, earthy, and acidic tend to come from Europe, while those from elsewhere more often feature riper, more dessert-like flavors, as well as more overt oak (see pp.162–173).

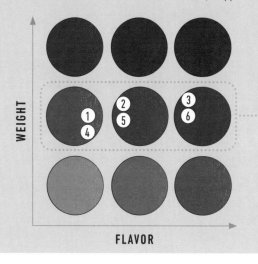

... TRY THESE
WINES WITH
SIMILAR SENSORY
PROFILES.

MILDER FLAVOR

MEDIUM FLAVOR

BOLDER FLAVOR

1 CHILEAN MERLOT

Merlot is often overshadowed by its famous cousin Cabernet Sauvignon. Softer and fruitier, it nonetheless makes some of the world's finest red wines. Merlot thrives in Chile's sunny climate, where it grows riper than in its native Bordeaux.

2 FRENCH CÔTES DU RHÔNE

Grenache grapes dominate the blend in these flavorful Mediterranean wines, providing their signature flavors of strawberries and white pepper. The region's stronger premium wines are named for their villages, like Châteauneuf-du-Pape.

3 FRENCH BORDEAUX

All but the mightiest red Bordeaux fall in the mid-weight category. It can be hard to tell whether they contain more Merlot or more Cabernet Sauvignon, but all are known for being lean, dry, and flavorful, with herbal aromatic accents.

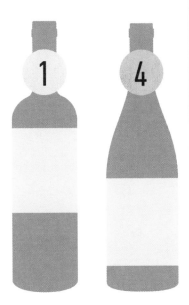

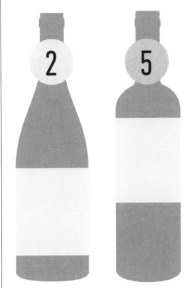

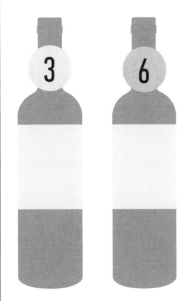

4 ITALIAN BARBERA

Barbera from Piedmont used to be known as a light, tart, fish-friendly red. Some are still made the traditional way, but many are now ripened longer and aged in barrels, resulting in a richer and more intense blackberry-scented wine.

5 ITALIAN MONTEPULCIANO D'ABRUZZO

Fans of Côtes du Rhône will find that these wines from Italy's Adriatic coast share many Rhône traits. They are modestly priced wines that deliver generous flavor and texture and a touch of rusticity, without feeling heavy.

6 SPANISH RIBERA DEL DUERO

Tempranillo is Spain's most noble grape. It ripens brilliantly on the arid plains of Castilla that line the Duero River, making some of Spain's most age-worthy red wines, which see French-style aging in new oak barrels.

EXPLORING HEAVIER RED STYLES

Red wines must reach higher alcoholic strength than whites to be heavyweight contenders: typically 14.5% alcohol or more. For standard wines, grapes must achieve uncommonly high ripeness to reach this level, from either very warm vineyards or by hanging extra-long on the vine. But this category also includes fortified red wines—wines whose alcoholic strength is enhanced with added brandy, such as strong, sweet Port.

IF YOU ENJOY THESE HEAVIER RED WINES ...

RIPE OLD AGE-WORTHY WINES

Heavyweight reds are naturally concentrated in flavor, and many are premium wines.

Since high-alcohol reds tend also to be high in tannin and flavor compounds, these wines often require time in barrels, and many are released years after the harvest. Flavors lean toward the black-fruit to baked-fruit range in this category, reflecting the taste of hyper-ripe fruit. Many wines feature spiced aromas such as pepper, cocoa, or clove, some derived from the grapes and others from maturation in new oak barrels.

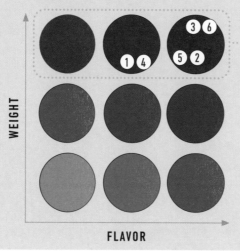

WEIGHT

FLAVOR

... TRY THESE WINES WITH SIMILAR SENSORY PROFILES.

MEDIUM FLAVOR

BOLDER FLAVOR

BOLDEST FLAVOR

1 PREMIUM AUSTRALIAN BAROSSA SHIRAZ

Thick-skinned Shiraz, aka Syrah, can make wines of epic intensity in regions like South Australia's Barossa and McLaren Vale. These inky wines have potent aromas of jam, bacon, and black pepper.

2 PREMIUM AMERICAN NAPA VALLEY CABERNET SAUVIGNON

Cabernet Sauvignon makes some of the world's most intense and long-lived red wines. The heaviest examples come from sun-drenched New World regions like California and feature flavors such as mocha and cassis liqueur.

3 PORTUGUESE PORT

The world's heaviest red wines are the fortified dessert wines of Portugal's Douro Valley. Spiked with distilled spirit mid-fermentation, they taste like a delicious mix of red wine, fresh grapes, and grappa.

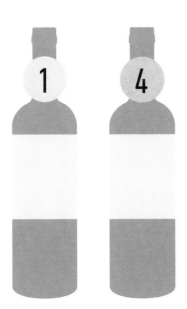 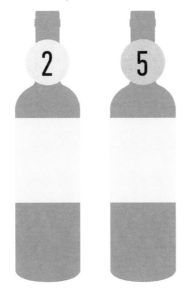 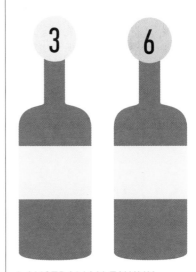

4 PREMIUM ARGENTINIAN MALBEC

Argentina's arid Mendoza plateau flanks the Andes at high elevation and makes intense red wines from the French Malbec grape. Premium bottlings can be blockbusters, with flavors of blueberries, five-spice, and violets.

5 PREMIUM SPANISH PRIORAT

This Catalan appellation is a treasure trove of ancient Garnacha and Cariñena vines whose wines have incredible concentration and power. Often blended with Syrah and Cabernet Sauvignon, they feature flavors of figs and anise.

6 AUSTRALIAN TAWNY

Australia, as a former British colony, was weaned on fortified Ports and Sherries and has a strong tradition of making delicious "stickies"—strong, sweet dessert wines. The term "Tawny" is used for an aged style with russet color and nutty, caramelized flavors.

MATCHING WINE AND FOOD

THE SOMMELIER'S SECRETS Wine is an excellent food partner, like a special sauce on the side, and most foods taste better with wine than with other drinks. For everyday meals, aligning your wine with the time of year or time of day is enough. But for real wine-and-food harmony, thinking like a sommelier can help you match recipes with flattering wines. Professionals know that the ways a dish are seasoned and cooked are often more important than its main ingredient. Discover some surprising quirks of sensory science that will change how you think about pairing wine and food.

WHAT TO DRINK WHEN

Outside of fine dining and fancy dinner parties, it's rarely necessary to worry about pairing specific wines to specific foods. Choices based on factors such as time of day and time of year often make more sense than trying to tailor a wine to a particular dish. The season affects what we're likely to be eating anyway, and most meals combine all sorts of diverse foods, whether served for lunch or dinner.

TOO HOT? CHILL OUT!

Wines that deliver refreshment and can be served chilled fit the bill in summer for the same reason we wear shorts and eat salads: They help us cool off.

OFFSET WEATHER WITH WINE

The same factor that has such a dramatic impact on wine style has a strong influence on our wine cravings. We instinctively seek lighter, younger wines (usually chilled) while the sun is up or the weather is hot. And when the sun goes down or the temperature drops, stronger, more complex wines make us feel warm and cozy. Drinking aged, full-bodied wines—particularly reds served at room temperature—is the beverage equivalent of pulling on a sweater. Essentially, bold, strong, warm-climate wines provide a little bottled sunshine that has the power to banish a wintry chill, while light, brisk cool-climate wines help us beat the heat.

WHEN THE SUN IS HIGH OR THE WEATHER IS WARM

WE TEND TO CRAVE WINES THAT ARE:

Lighter
Lower in alcohol
Younger and fresher-tasting
Lower in oak flavor
Served colder
Paler in color

Simple, young, sparkling wines,
like Italian Prosecco

Crisp, unoaked whites,
like New Zealand Sauvignon Blanc

Light, refreshing "chillable" reds,
like French Beaujolais

Pale, brisk, fortified wines,
like Spanish Manzanilla Sherry

WHEN THE SUN GOES DOWN OR THE WEATHER IS COOL

WE TEND TO CRAVE WINES THAT ARE:

Heavier
Higher in alcohol
Mature and complex-tasting
Higher in oak flavor
Served warmer
Deeper in color

Complex, aged, sparkling wines,
like French Champagne

Rich, barrel-fermented whites,
like California Chardonnay

Strong, flavorful, "warming" reds,
like Australian Shiraz

Opulent, jewel-toned, fortified wines,
like Portuguese vintage-style Port

MATCHING ON MANY LEVELS

Most wines taste great with most foods, so it's hard to go too far wrong: Taking cues from the season, time of day, and degree of formality is usually all that's necessary to improve wine and food harmony. However, when the occasion calls for something special, borrow a few pairing strategies from wine professionals to tilt the odds in your favor.

Consider which types of wines offer similar flavors or textures and so might harmonize with your meal's main ingredient. Paying attention to seasoning and cooking methods allows sommeliers to create pairings in which both the wine and food magically seem to taste better together than they did apart. You don't need encyclopedic wine knowledge to get there; all you need is a little insight into how our senses work when wine and food interact.

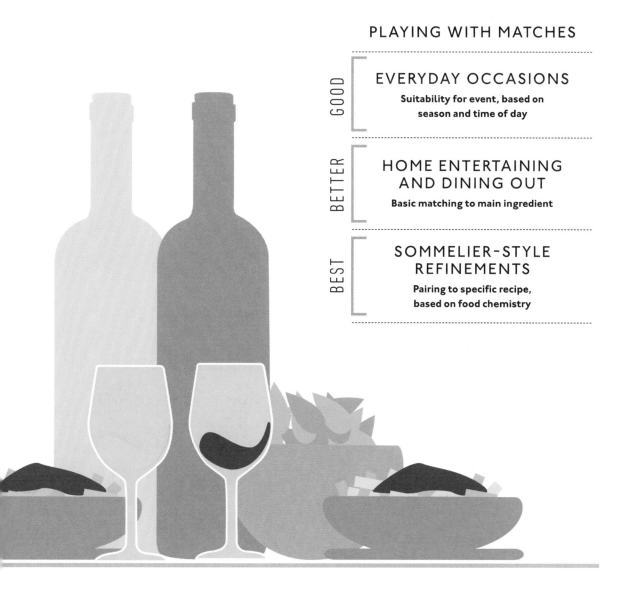

PLAYING WITH MATCHES

GOOD

EVERYDAY OCCASIONS
Suitability for event, based on season and time of day

BETTER

HOME ENTERTAINING AND DINING OUT
Basic matching to main ingredient

BEST

SOMMELIER-STYLE REFINEMENTS
Pairing to specific recipe, based on food chemistry

PAIRING WINE WITH MAIN INGREDIENTS

When choosing a specific wine to go with a certain meal, most people have excellent natural instincts. Successful wine pairings are based on the same principles that lead us to drink lemonade with summer salads or espresso with chocolate desserts. Pairing wines that share similar sensory qualities with the food being served—light with light, rich with rich—is a very effective way to find mutually flattering combinations.

MATCH WEIGHT AND TEXTURE

The lightest foods, like delicate shellfish, usually taste best with the lightest wines, such as sparkling whites, while the heaviest foods, like red meats, usually taste best with stronger, heavier wines, like intense reds. Both fat in food and alcohol in wine register as texture in the mouth; fatty meats and high-alcohol wines both seem thicker and heavier than low-fat seafood and low-alcohol wines.

SURF OR TURF?
Vintners often tailor their wine styles to flatter local cuisine. A delicate seafood dish is unlikely to match well with a bold red from cattle country, so consider a subtle coastal white.

WHY IS MATCHING FLATTERING?

When we drink wine with our dinner, it's a balancing act in which the food and wine both compete for our attention. Sticking to drinks of similar weight and flavor density levels the playing field, like organizing wrestlers by weight class. When one is heavier or tastes stronger than the other, it can detract from our appreciation of the lighter partner, and the dining experience as a whole suffers. The ideal is for neither element to overpower the other, allowing their flavors to mix and mingle on equal terms.

MATCH FLAVOR INTENSITY

Understated foods with mild flavor, like oysters or omelets, generally taste best with mild white wines, such as Muscadet or Pinot Grigio. Foods that are higher in flavor intensity, like smoked salmon or blue cheese, tend to taste better with wines of similar aromatic amplitude, like Gewurztraminer or Cabernet Sauvignon.

MATCH COLOR DEPTH

Absence of color tends to signal low flavor intensity, while deep color tends to coincide with high flavor impact (though there are exceptions, of course). Think of the difference in flavor power between flounder and tuna, chicken and duck, or veal and venison. Foods with minimal color, like goat cheese or scallops, tend to taste best with transparent white or sparkling wines, while foods with intense color saturation, like lamb or chocolate, tend to taste better with red wines.

MILD FLAVORS
SUSHI; NOODLES

BOLD FLAVORS
SAUSAGES; CURRIES

LIGHT COLORS
GOAT CHEESE; SCALLOPS

STRONG COLORS
LAMB; CHOCOLATE

THINK FLAVOR; THINK COLOR
Consider the main ingredients, and think about which wine styles might have a broad resemblance in terms of weight, flavor intensity, or color.

PAIRING TRICKS OF THE TRADE

MAPPING FOODS AND WINES BY WEIGHT AND FLAVOR INTENSITY CAN HELP YOU IDENTIFY THOSE WITH SIMILAR QUALITIES.

Compared to wine, the foods we eat are far more complicated: More ingredients and recipe combinations lead to more possibilities. If we want to choose a wine for a particular meal, it helps to understand how foods relate to one another in weight and flavor intensity, in the same way that we assess wine styles.

ASSESS HOW FOODS RELATE

Thinking about how foods relate to one another allows us to predict which wines will best suit any given meal. Between core proteins and side dishes, sauces and garnishes, though, it can be hard to know where to focus. And in a restaurant, everyone at your table may be eating a different dish, so it helps to think in terms of dominant features and cuisine styles.

LIGHTER OR HEAVIER?

When we look at menu items in restaurants or recipes in cookbooks, we instinctively consider their "weight." We might want something light, like a salad, or something heavier, like a steak. Regardless of portion size, fat content is the main feature that drives perceived weight in food. The mouth-coating richness of fats and oils gives food added texture in the same way alcohol makes wine feel heavier.

MILDER OR BOLDER?

Flavor intensity is something we think about less often, but it plays a major role in personal taste. Some foods are inherently more flavorful than others. Tomatoes have a stronger taste than cucumbers, for example, while chicken tastes milder than lamb.

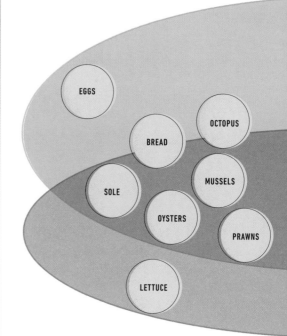

LIGHTER OR HEAVIER
Fat content and protein density are the main variables that make foods seem more delicate or more rich.

FACTOR IN THE RECIPE

Cooking methods and seasonings often boost flavor and enrich texture, which on the chart below can push any of the items higher or farther to the right. (The items on the chart are only examples, and it should not be considered an exhaustive list.) Chicken will feel heavier if fried, and bolder if grilled, than it would if it were steamed. The seasonings, sauces, and marinades associated with Thai cooking consistently add a spicy-sweet flavor boost, regardless of whether they are added to shrimp or beef.

WHITE WINE
like Chardonnay
and Pinot Grigio

SPARKLING WINE
like Champagne
and Prosecco

ROSÉ WINE
like Anjou
and Tavel

RED WINE
like Shiraz
and Chianti

FORTIFIED WINE
like Port
and Sherry

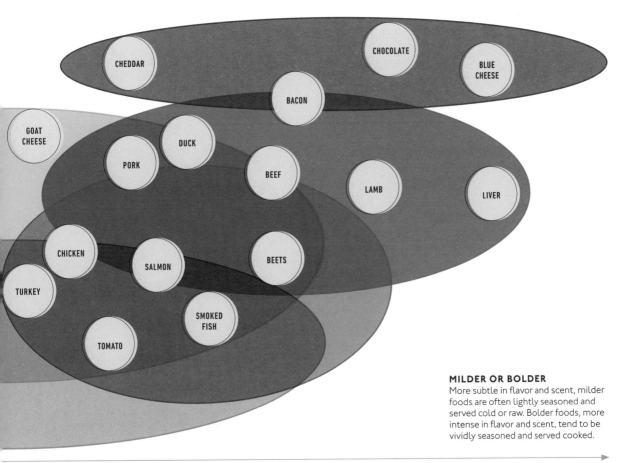

MILDER OR BOLDER
More subtle in flavor and scent, milder foods are often lightly seasoned and served cold or raw. Bolder foods, more intense in flavor and scent, tend to be vividly seasoned and served cooked.

FLAVOR

BOLDER

PAIRING WINE TO SPECIFIC RECIPES

While matching by main ingredient works reasonably well, it is possible to take your wine-pairing skills to the next level by learning a few tricks of the trade. Professional sommeliers almost always rely on the "like with like" strategy for making pairing recommendations, but they achieve better results because they take it a step further.

PRO MATCHING

When sommeliers match wine and food, they assess the total sensory experience of each dish, including how it's cooked, sauced, and seasoned.

MATCH TO STRONGEST FLAVORS

Wine drinkers usually work around main ingredients when choosing wine—for example, picking a Pinot Grigio for scallops or a Cabernet Sauvignon for beef. Sommeliers pay more attention to overall flavor. If the recipe is tangy or herbal or smoky or sweet, they will generally choose wines with similar flavor profiles, regardless of the main ingredient. They look beyond the protein at the heart of a dish and consider how the recipe makes it look, taste, smell, and feel in the mouth.

EVERYDAY MATCHING BY INGREDIENT

Scallops are delicate and pale. *Pair them with delicate white wines, like Italian Pinot Grigio.*

Beef is a stronger dark-red meat. *Pair it with heavier red wines, like California Cabernet Sauvignon.*

SOMMELIER-STYLE MATCHING BY PREPARATION

Scallop ceviche is tangy, citrusy, and herbal. *Pair it with a tangy, citrusy, herbal white wine, like New Zealand Sauvignon Blanc.*

Beef carpaccio is raw and barely seasoned. *Pair it with a young, refreshing white wine with subtle flavor, like Spanish Albariño.*

Seared truffled scallops are rich, earthy, and caramelized. *Pair them with a rich, earthy, and oaky white wine, like French Meursault (Chardonnay).*

Steak au poivre is dark, intense, and peppery. *Pair it with an inky-purple, black-pepper-scented red wine, like Australian Shiraz.*

MATCH OAKED WINES TO BROWNED FOODS

The distinctive Cognac-like flavor of new oak in wine has a strong similarity to the flavors found in foods that have been browned during cooking. Roasting, frying, or grilling a chicken breast will deepen its color and intensify its flavor in a way that steaming or boiling will not. Aging or fermenting wine in new oak barrels can add a similar set of flavors and scents, since wood is "toasted" in the barrel-making process. The aromatic resemblance between browned foods and barrel-aged or barrel-fermented wine provides them with a natural affinity.

LIGHTEN UP A LITTLE

In restaurants, wine is not the star of the show. Wines are designed to work in a supporting role to highlight the food. Just as background harmonies shouldn't obscure a lead singer's melody, wine shouldn't interfere with our ability to appreciate a chef's handiwork.

Lighter, milder wines are more forgiving food partners. Not only do they pair pleasantly with lighter, milder foods, but they can also provide a palate-cleansing counterpoint even when outmatched by heavier meals. Strong, intense wines may taste impressive alone, but their forceful traits can get in the way when paired with food, unless the dish is equally hearty and flavorful.

▼ BUBBLES ARE A SAFE BET
If you are looking for a wine that will go with anything, something lighter and milder is the safest choice. For sommeliers, the rule of thumb is, "When in doubt, pair Champagne."

▲ COLOR MATCHING
In general, foods that are raw or cooked without browning pair best with wines that are young, unoaked, and fresh. Foods that are browned in cooking—roasted, grilled, or sautéed, for example—tend to be better with oakier wines.

FOOD CHEMISTRY

Sommeliers know that seasonings in food can skew our sensory perceptions of wine in predictable ways. Salt and sugar in particular can significantly alter how we perceive the tastes of acidity and sweetness in any wine served alongside our meals.

SALT: DRY WINE'S FRIEND

Salt is present in most things we eat, and we tend to add more in cooking because it helps emphasize other flavors. Salt blocks our ability to discern acidity in wine, and wine returns the favor by blocking perceived saltiness in the food without compromising overall flavor.

Wines need to taste a little too sour alone to taste just right with most foods. In fact, one reason many wines are so acidic is that winemakers know they will be drunk with salted food. Tart wines designed around this principle—like Italian Chianti or French Sancerre—are often called "food-oriented" wines.

SUGAR: DRY WINE'S ENEMY

Also a universal flavor enhancer, sugar is less wine-friendly than salt. Sweet food makes wine taste far more sour than it seems alone. Since wine is naturally tart, the result is usually unflattering, like the shocking acidity of orange juice after brushing your teeth.

A wine needs to be at least as sweet as a dish to avoid this problem. Foods with faint sweetness are far kinder to fruity modern wines than to drier styles. Overt sweetness, as found in fruit-based sauces, works well with lightly sweet wines like Riesling. Fully sweet desserts call for sticky-sweet dessert wines.

SALT IN FOOD REDUCES WINE'S PERCEIVED ACIDITY.

SUGAR IN FOOD INCREASES WINE'S PERCEIVED ACIDITY.

LESS SOUR · MORE SOUR

THE TASTING

Identifying the Effects of Salt and Sugar

COMPARE TWO WINES AT HOME

Prepare a glass of each of the wine styles below.

1 Sample the wines, paying special attention to first impressions of sweetness and acidity.
2 Taste a pinch of salt. After it dissolves, taste the Sauvignon Blanc again. Note how the perception of acidity is significantly reduced, making the wine seem softer and fruitier.
3 Repeat step 2 with the Riesling. Notice how the same acid-blocking effect takes place and that suppressing acidity makes it seem a little sweeter.

4 Wait 5 minutes to allow your taste buds time to recover, then repeat step 1.
5 Taste a dab of honey, and sample the Sauvignon Blanc again. Note how the acidity is amplified, making the wine seem extremely sour and one-dimensional.
6 Repeat step 5 with the Riesling. Notice how the same acid-boosting effect takes place, but is less unpleasant with the sweeter wine. The wine now seems significantly less sweet than it did alone or after salt. The honey's sweetness overpowers the wine's, making it seem much drier.

FRENCH SAUVIGNON BLANC

AMERICAN RIESLING

For example ... Bordeaux Blanc, Sancerre, Pouilly-Fumé, or Touraine

By itself, notice ... Low sweetness/very dry; high acidity/very tart

After salt, notice ... Marginally less dry; dramatically less acidic

After honey, notice ... Marginally less sweet; dramatically more acidic

For example ... Styles from Washington State's Columbia Valley or California that do not say "dry" on label

By itself, notice ... Medium sweetness/lightly sweet; medium acidity/tangy

After salt, notice ... Marginally sweeter; dramatically less acidic

After honey, notice ... Dramatically drier/less sweet; dramatically more acidic

BEYOND SALT AND SUGAR: HOW YOUR SENSES ADJUST

Most sensations don't "add up" to seem stronger together; they balance and neutralize each other as our senses adjust. Sights and sounds seem most vivid when isolated, and any competing stimulus reduces their perceived intensity—that's why we can hear a whisper in a quiet room but must shout to be heard in noisy restaurants. The same holds true for most tastes, smells, and textures, but we are much less likely to notice the pattern.

This effect is usually most obvious and immediate with taste-bud sensations.

When wine and food share a similar sensory characteristic, your senses adjust and perceive both as weaker together than they were apart. The result is nearly always pleasant and harmonious, allowing us to apply the broad "like with like" approach to tastes, smells, and textures across the board.

Not only do we pair light with light, and heavy with heavy, but we can also pair tart with tart, oaky with smoky, and sweet with sweet, with terrific results.

IN SENSORY PERCEPTION, ONE PLUS ONE DOES NOT EQUAL TWO!

Tangy foods—tomatoes or pickles, say—and tart wines—like Chianti or Sauvignon Blanc—will seem less acidic together, not more acidic.

Sweet foods—such as fruit salad or crème brûlée—and sweet wines—like Riesling or Sauternes—will seem less sweet together, not more sugary.

A similar pattern occurs with many olfactory and tactile perceptions, too, though less vividly:

Smoky foods—smoked salmon, grilled meats—and oaky wines—Chardonnay or Rioja—will seem less woody together, not more woody.

Rich foods—like truffled risotto or chocolate mousse—and high-alcohol, full-bodied wines—Barolo or Port—will seem *lighter* together, not heavier.

WE DON'T TURN ON THE LIGHTS TO SEE THE TV BETTER, BECAUSE TWO COMPETING LIGHT SOURCES MAKE EACH OTHER SEEM DIMMER. A SIMILAR NEUTRALIZING EFFECT TAKES PLACE WHEN WINE AND FOOD SHARE A DOMINANT FEATURE.

FINE DINING IS NOT AN EXTREME SPORT

Sommeliers often recommend wines that share a dominant feature with the food that is being served. The goal is to deliberately allow these strong features to tone each other down—that is, not to exaggerate the shared trait but to help it melt into the background in a pleasing way. Pairing wine and food is not like watching movies or going to rock concerts, where isolating and amplifying sensations can push the envelope to thrilling extremes. Harmony between wine and food is almost always about seeking comfort and balance.

SPICY-HEAT EXCEPTION

"Like with like" is a sound pairing strategy, but it doesn't work for spicy foods and "spicy" wines. The same word may be used for both, but it describes very different qualities. Foods that are spicy trigger a physical sensation of burning, like that of hot chiles, whereas wines that are spicy have intense aromatics that resemble spices or seasonings, like those of peppery Syrah. Since wine aromatics come mainly from grape skins and intensify with ripeness, the wines with the spiciest flavors are most often red and full-bodied.

Sommeliers tend to avoid heavier reds with spicy heat and, more often, recommend lightly sweet white and rosé wines with low alcohol—an Italian Moscato or German Riesling, for example.

Spicy heat in food and high alcohol in wine amplify each other's dominant features instead of balancing and neutralizing. Alcohol acts as an irritant that briefly makes the burn of spicy food seem more intense and painful, like rubbing salt into a wound. And full-bodied wines seem heavier and more alcoholic when tasted after eating something spicy. Neither effect is flattering.

DON'T BURN YOUR TONGUE!

The high alcohol level of heavy wines can make them painful to drink with foods that have a fiery kick, so avoid pairing them with the hottest spicy dishes.

UPPING THE ANTE
Low-alcohol wines tame the "flames" of spicy food and soothe the senses, while heavier wines seem to turn up the heat.

SPICY HEAT IN FOOD

HIGH ALCOHOL IN WINE TURNS UP THE HEAT

THE TASTING

Identifying Sensory Competition

COMPARE FOUR WINES AND FOUR FOODS AT HOME

Try this simple taste test at home to get a clear idea of how your senses operate when they encounter competing sources of similar sensation.

1 Sample all the wines alone first, paying special attention to your first impressions of their sweetness and acidity, fruit and oak flavor, and body and tannin.

2 Take a bite of fresh tomato, then retaste wine 1. Notice how eating something equally tangy makes the wine seem less tart than it did alone, not more acidic.

3 Retaste wine 2; then take a bite of smoked almond, and taste it again. Notice how eating something roasted or smoked with an equally woody flavor makes the wine seem less oaky than it did alone, not more so.

4 Retaste wine 3 both before and after a dab of butter. Notice how tasting something equally rich and heavy makes the wine seem lighter than it did alone, not heavier.

1 YOUNG ITALIAN SANGIOVESE

2 TEMPRANILLO-BASED SPANISH RESERVA

3 PREMIUM NEW WORLD CABERNET SAUVIGNON

For example ... Chianti Classico or any young, modestly priced Tuscan Sangiovese-based wine that is less than three years old

On its own, you should detect ...
Low sweetness/very dry;
high acidity/very tart;
medium fruit intensity;
low oak flavor;
mid-weight;
medium tannin

Make it less acidic
Acidic foods like tangy tomatoes, citrus, or vinegar will make wine tasted alongside seem dramatically less acidic.

For example ... Rioja Reserva or any *reserva*-level Spanish Tempranillo-based wine, such as Ribera del Duero or Toro

On its own, you should detect ...
Low sweetness/dry;
high acidity/tart;
medium fruit intensity;
High oak flavor;
mid-weight;
medium tannin

Make it less oaky
Smoked almonds or other foods with roasted, browned flavors will make wine tasted alongside seem much less oaky.

For example ... Chilean, or any premium Cabernet Sauvignon–based wine from the Americas or southern hemisphere

On its own, you should detect ...
Low sweetness/dry;
medium acidity/tangy;
high fruit intensity;
high oak flavor;
heavy;
high tannin

Make it less heavy
Fatty foods like dairy products or meats will make wine tasted alongside seem noticeably lighter.

5 Retaste wine 4 before and after a bite of dark chocolate. Notice how eating something equally sweet makes the wine seem less sweet than it did alone, not more sugary.
6 Wait 5 minutes to allow your taste buds to recover from all this stimulation.
7 Retaste any combination of wine and food in the same before-and-after format. You should notice two things: a) the effects of tongue-based tastes, like sweetness and acidity, are more dramatic than olfactory and tactile sensations; and b) the food items generally seem most harmonious with the wines that most resemble them in taste, smell, or texture.

For example ... Portuguese Port or any sweet, fortified dessert red wine over 15% alcohol, such as Australian Tawny or French Banyuls

On its own, you should detect ...
Very high sweetness;
low acid/soft;
very high fruit intensity;
variable oak by style;
fortified/very heavy;
medium tannin

Make it less sweet
Sweet foods like candy, desserts, or fruit will make wine tasted alongside seem dramatically less sweet and more acidic.

CHAPTER CHECKLIST
Here is a recap of some of the most important points you've learned in this chapter.

- Don't worry about pairing **specific wines to specific foods**. It is often better to make choices based on such factors as **time of day and time of year**.
- **Lighter, younger wines** (usually chilled) are ideal in hot weather, while **stronger, more complex wines** are perfect for cooler seasons.
- The most flattering combinations come when you pair wines with **similar sensory qualities** as the food being served: light with light, and rich with rich.
- Match wine and food according to **intensity of flavor or depth of color**, and you'll get it right more often than you might expect.
- **Seasonings and sauces** often influence a dish's flavor more than the main ingredients, and this is key to how sommeliers work, assessing the **total sensory experience** of each dish—even down to the cooking methods used—to find the best wine match.
- Generally speaking, raw foods and foods cooked without browning match best with **young, fresh, unoaked wines**. Browned foods— roasted, grilled, or sautéed, for example—tend to work better with **oakier wines**.
- When in doubt, pair **Champagne** or **sparkling wine**.
- **Salt and sugar** in a dish alter your perception of acidity and sweetness in any **accompanying wine**. Salt in food reduces wine's perceived acidity; sugar increases wine's perceived acidity.
- **"Like with like"** is a sound pairing strategy; however, spicy foods don't work with so-called spicy wines. These **big, bold reds** seem to magnify the **fiery burn** of a dish, while low-alcohol white wines **soothe** the senses.

MASTERING WINE VARIABLES

WINE IS AN INCREDIBLY complicated consumer product, even when we have a few rules to guide us. Wine labels emphasize grapes and regions, brands and vintages, but none of these spells out how a wine will taste. It can feel like they're written in code. Wine beginners often wish they knew more about each wine—the proportions of the blend, or the number of months it ages in barrels—but that's rarely helpful information. Experts know that a short list of variables shapes wine's flavor, such as vineyard climate and winemaking decisions. This lets them make useful generalizations. For the wine novice, it's more fruitful to get acquainted with how these controlling factors work than it is to get bogged down in brand-specific details. There will always be exceptions, of course, but discovering tasty surprises is what keeps wine interesting.

WINEMAKING DECISIONS

SWEETNESS, COLOR, AND OAK A vineyard's climate may control its potential for ripening fruit, and grape varieties may have distinct flavor profiles, but these are only wine's raw materials. Human decisions at the winery also play a major role in a wine's flavor. Like a chef given fresh ingredients, the winemaker decides what to do with them—whether to make sweet wine or dry, red wine or white, wine that is fresh and unoaked or barrel-aged and refined. By manipulating the winemaking process, he or she can also turn the grapes into sparkling wine with bubbles or fortified wine with extra alcohol.

FERMENTING GRAPES INTO WINE

Wine is produced through fermentation, a process in which living yeasts convert sugar into alcohol. In fresh foods, this natural occurrence is the first step on the road to spoilage. However, for thousands of years, humankind has taken control of the fermentation process—and not only for making wine and beer. Fermentation turns flour into bread, milk into cheese, and cocoa beans into chocolate. Think of yeast as the magic pixie dust that transforms simple fresh grapes into much more complex, flavorful wine.

◄ **WHAT ARE YEASTS?**
Yeasts are microscopic organisms, single-celled members of the fungus kingdom.

SUGAR BREAKDOWN

Many types of yeasts occur naturally in our environment, particularly those of the sugar-eating *Saccharomyces* genus that is used in baking, brewing, and winemaking. These yeasts consume sugar and break it down into alcohol and carbon dioxide (CO_2). Fermentation always begins spontaneously, thanks to the presence of wild yeasts in the vineyard and winery. Nowadays, though, many modern vintners prefer to inoculate with cultured yeast strains for more predictable results.

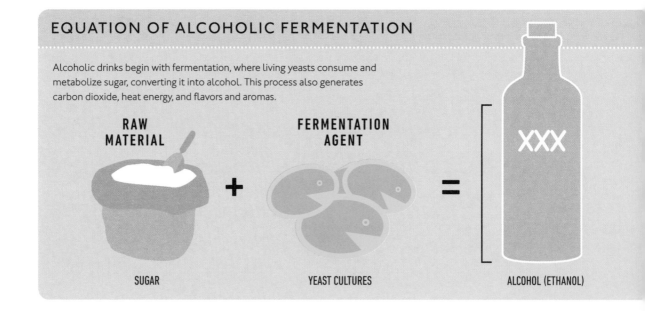

EQUATION OF ALCOHOLIC FERMENTATION

Alcoholic drinks begin with fermentation, where living yeasts consume and metabolize sugar, converting it into alcohol. This process also generates carbon dioxide, heat energy, and flavors and aromas.

RAW MATERIAL

+

FERMENTATION AGENT

=

XXX

SUGAR

YEAST CULTURES

ALCOHOL (ETHANOL)

FLAVORS FROM BEYOND THE GRAPE

Wines smell and taste of far more than just grapes. During fermentation, countless smaller chemical reactions are triggered, adding new flavors and scents that may not have been present or detectable in the fruit.

Fermentation is what makes wine so aromatically complex and enjoyable, much the way it does for cheeses. The unique flavors of French Brie, Wisconsin Cheddar, and Italian Gorgonzola derive from fermentation and aging, not from their raw material (cow's milk). Yeast significantly alters the flavors and scents found in fresh grapes, too.

CHOCOLATY OR PEPPERY WINE?

Exotic terms are often used to describe non-grape aromatics in wine, usually by naming other fruits, foods, or spices they resemble. These are not ingredients, though; they are simply metaphors to describe wine's diverse scents.

▶ **IMPORTANCE OF CULTURES**
Just like cheese makers, vintners choose their yeast strains carefully to exert control over how the finished wine will taste and smell.

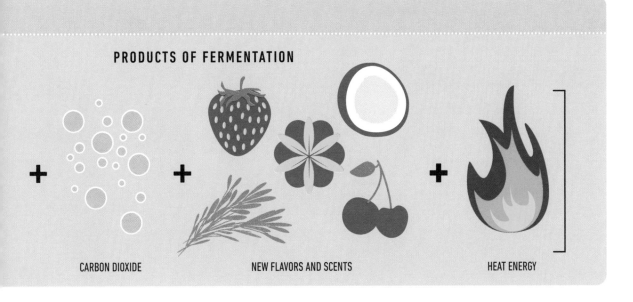

PRODUCTS OF FERMENTATION

+ CARBON DIOXIDE

+ NEW FLAVORS AND SCENTS

+ HEAT ENERGY

CONTROLLING SWEETNESS

Grapes are sweet, but most wines are not, and the historic reason for this is entirely practical: early winemakers wanted to increase alcohol and reduce sugar to prevent spoilage. Sweet wines and low-alcohol wines are most susceptible to microbiological decay, whereas more alcoholic wines that are drier—that is, less sweet—have a longer shelf life.

WHY SO DRY?

Wines are traditionally fermented dry because this has been the easiest way to make wine for thousands of years. Once fermentation begins, yeasts feed and reproduce until their sugar supply is depleted, and it is difficult to halt the process prematurely.

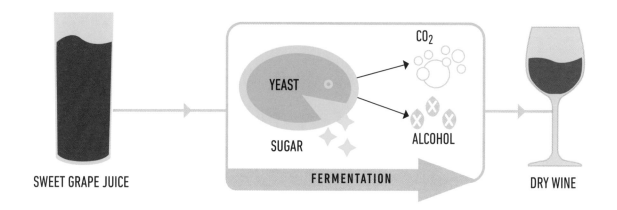

SWEET GRAPE JUICE

YEAST

SUGAR

CO_2

ALCOHOL

FERMENTATION

DRY WINE

SUGAR IS POTENTIAL ALCOHOL
Sugar's primary role in winemaking is to be converted into alcohol. Most wines are dry, not sweet, especially those with more than 13% alcohol.

Lightly sweet wines

Fully dry wines

18% sugar

0% alcohol

10% alcohol

3% sugar

13.5% alcohol

0.5% sugar

SUGAR (%)

ALCOHOL (%)

SWEET WINEMAKING DREAMS

Sweet wines may be rare, but they have always been desirable because they taste delicious. Most European wine regions have developed their own methods for making sweet wines, each an adaptation to their environment. Modern technology now makes these wines more stable than ever. Most follow one of three winemaking strategies, each of which can make wines that range from lightly sweet "off-dry" wines to candy-sweet dessert wines.

METHOD 1: INTERVENE TO STOP FERMENTATION EARLY

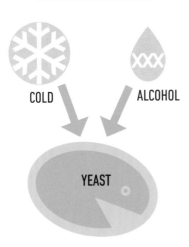

COLD   ALCOHOL

YEAST

Interrupting the yeast life cycle can preserve some natural grape sugar. There are two main options.

1 Reduce temperature
Yeasts slow and die in near-freezing conditions. Retaining sugar this way means sacrificing potential alcohol, so the sweetest wines will have the lowest alcohol content.
Examples: *lightly sweet Rieslings and rosés; fully sweet Italian Moscatos*

2 Add distilled spirits
Yeasts cannot tolerate alcohol over 15%, so fortifying wine with brandy can stop fermentation.
Examples: *fully sweet Portuguese Ports; French Vin Doux Naturel Muscats*

METHOD 2: CONCENTRATE GRAPES BEFORE FERMENTING

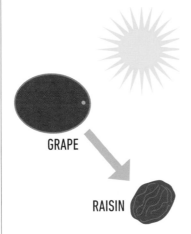

GRAPE

RAISIN

Reducing the water content of grapes increases the proportion of all that remains, including sugars, acids, and flavor compounds. In warm regions, grapes may be sun-dried after the harvest; in cooler zones, delaying harvest is more common. "Late-harvest" grapes shrivel and concentrate on the vine, yielding rich, sweet wines. In cold climates, grapes may hang until midwinter, freeze-concentrating remaining juice.
Examples: *sun-dried Italian Vin Santo and Spanish Moscatel; late-harvest French Sauternes and German Auslese; freeze-concentrated Austrian Eiswein and Canadian Icewine*

METHOD 3: SWEETEN THE WINE AFTER FERMENTING

SUGAR

WINE

Many bargain wines that aren't fully dry are made by blending small amounts of grape juice (or concentrate) into a dry wine. Many fine sweet wines are also made by variations on this theme. Most use grapes as their sugar source, but some—like *demi-sec* Champagne—use cane sugar.
Examples: *German Liebfraumilch (grape juice); French Champagne (cane sugar); Spanish Cream Sherry (raisin syrup); Hungarian Tokaji Aszú (late-harvest grapes)*

THE TASTING

Identifying Stages of Fermentation

COMPARE GRAPE JUICE AND WINE AT HOME

All wines start as sweet grape juice, such as example 1, and most end up as dry wines with no perceptible sugar, but plenty of alcohol, like example 4. However, winemakers can interrupt the fermentation process to make a sweeter wine. Wines 2 and 3 are made by disrupting fermentation.

1 Sample these four drinks side by side. As you do so, consider how fermenting with yeast changes grape juice into wine.

2 Pay particular attention to the palpable variations in sweetness and body as the wines grow heavier in weight.

3 Notice that the presence or absence of sugar and alcohol are not obvious when smelling but become very apparent on the palate when you take a sip.

MOST WINES ARE
FERMENTED FULLY DRY, BUT
VINTNERS SOMETIMES STOP
FERMENTATION EARLY TO
RETAIN GRAPEY SWEETNESS.

Sugar

1

WHITE GRAPE JUICE

For example ... Any commercial juice, or squeeze your own

Contains
100% table grapes
18% sugar (approx)
0% alcohol (approx)

Can you detect ...?
Very sweet;
very tart;
grapey scents and flavors

Explanation Grape juice is the sweetest of all fruit juices, which is why it is ideal for winemaking. Wines made from other fruits are lower in alcohol and have a shorter shelf life. Wine is rarely made from the kinds of table grapes used for juice and jelly, but they are similar enough in sweetness to help us picture the raw material vintners use to make white wines.

Ferment halfway; stop by chilling	Ferment halfway; stop by adding brandy	Ferment completely, depleting all sugar

For example ... Italian Asti or any sweet, sparkling Moscato that contains less than 10% alcohol

Contains
100% Moscato (aka Muscat)
5% sugar (approx)
7% alcohol (approx)

Can you detect ...?
High sugar/fully sweet;
high acidity/tart;
high fruit intensity;
no oak flavor;
low alcohol/light;
high carbonation

Explanation Fizzy, sweet Moscatos taste like what they are: a halfway point between grape juice and wine. Such wines are made by stopping fermentation before it is complete, leaving lots of grape sugar unfermented and sacrificing potential alcohol. Their bubbles are a natural by-product of fermentation.

For example ... Muscat de Beaumes-de-Venise, Muscat de Minervois, or Muscat de Frontignan

Contains
100% Moscato (aka Muscat)
15% sugar (approx)
15% alcohol (approx)

Can you detect ...?
High sugar/very sweet;
low acidity/not refreshing;
high fruit intensity;
no oak flavor;
high alcohol/heavy;
no carbonation

Explanation This is a fully sweet style in which fermentation is interrupted before grape sugars are depleted—but through a different method. Adding distilled spirits mid-fermentation kills off the yeasts, creating a sweet, strong liqueur-like dessert wine.

For example ... French Alsace Muscat or any Muscat labeled "dry" from Australia or the US. Alternately, Argentinian Torrontés. (Do not choose French Muscats labeled *vendange tardive, sélection de grains noble,* or *vin doux naturel.*)

Contains
100% Moscato (aka Muscat)
0.5% sugar (approx)
13% alcohol (approx)

Can you detect ...?
Low sugar/dry;
high acidity/tart;
high fruit intensity;
no oak flavor;
medium alcohol/mid-weight;
no carbonation

Explanation Fermented to dryness, leaving no significant residual sugar, most dry wines do not taste at all sweet. They usually have at least 12% alcohol, which prevents spoilage.

DETERMINING COLOR AND STYLE

White and red wines are typically made with different types of grapes, but this isn't the source of the biggest differences in how they taste. White and red wines are made using two completely different processes.

TWO WINES, TWO METHODS

Consider following two recipes with the same ingredients: tomatoes, onions, and peppers. Peeling the tomatoes and chilling the mixture creates a delicate gazpacho soup that retains a garden-fresh taste without any bitterness from the skins. Keeping the tomato skins on and simmering the mixture changes the result dramatically, yielding a thicker, stronger-tasting pasta sauce. Grapes behave similarly. Making white wine involves eliminating grape skins and preserving the fresh taste of the juice through refrigeration; making red wine involves grape-skin contact and controlled heat to extract as much color and flavor as possible.

THE RED METHOD

RED WINES TASTE STRONGER AND MORE "BITTER," LIKE THE PEELS OF GRAPES.

This is because red wines are made from whole grapes, including skins, seeds, and pulp. Red wine can only be made from dark-skinned grapes, because the skins are their source of color and flavor.

THE WHITE METHOD

WHITE WINES TASTE MILDER AND MORE "JUICY," LIKE THE FLESH OF GRAPES.

This is because white wines are made from grape juice only—skins, seeds, and pulp are all discarded. Only clear juice is used in white-wine making. All grape solids, including the colorful skins, are removed before fermentation begins.

REDS USUALLY FERMENT FOR 1–3 WEEKS

WARM, FAST FERMENTATION EXTRACTS COLOR AND FLAVOR FROM DARK GRAPE SKINS

WHITES USUALLY FERMENT FOR 2–6 WEEKS

COLD, SLOW FERMENTATION PRESERVES FRESH TASTE OF CLEAR GRAPE JUICE

FERMENTED WARM AND FAST

Vintners harness the heat generated during fermentation to help extract color and flavor compounds from grape skins, and astringent antioxidant tannins come along for the ride. Warmth speeds the yeast life cycle, so red wines ferment to dryness quickly and vigorously, generating more of the chemical reactions that create new flavors and scents that weren't present in the grapes.

FERMENTED COLD AND SLOW

Winemakers use refrigeration to preserve the delicate flavor of the milder grape juice. Without grape-skin antioxidants present, they must also shield the juice from air in sealed tanks or barrels to prevent oxidation. Chilling slows the metabolism of yeasts, so fermentation proceeds much more gently. The rate of chemical reactions that generate new aromatic compounds is significantly reduced.

REDS, WHITES ... AND A BIT OF BOTH

A critical distinction between wine styles is when their grapes are pressed, separating liquid from solids: before fermentation for whites, after fermentation for reds, and mid-fermentation for rosés.

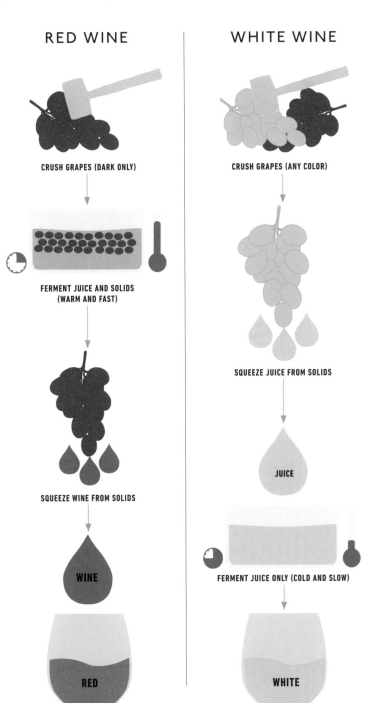

RED WINE

CRUSH GRAPES (DARK ONLY)

FERMENT JUICE AND SOLIDS (WARM AND FAST)

SQUEEZE WINE FROM SOLIDS

WINE

RED

WHITE WINE

CRUSH GRAPES (ANY COLOR)

SQUEEZE JUICE FROM SOLIDS

JUICE

FERMENT JUICE ONLY (COLD AND SLOW)

WHITE

ROSÉ WINE

CRUSH GRAPES (DARK ONLY)

FERMENT FOR UP TO 48 HOURS

SQUEEZE JUICE FROM SOLIDS

JUICE

FERMENT JUICE ONLY (COLD AND SLOW)

PINK

FERMENTING OR AGING IN OAK

Wines were once made and sold in barrels, but nowadays the vast majority are fermented in inert stainless-steel tanks and sold in bottles. However, winemakers still use traditional oak barrels to refine and improve most premium wines. In the same way that chefs use butter and spices to enrich and season recipes, vintners use oak barrels to add texture and flavor to their wines.

THE BARREL EFFECTS

There are three ways in which wine is changed by the time it spends in oak barrels.

ALL BARRELS INTENSIFY WINE
Water and alcohol are absorbed through wood's pores and evaporate when they reach the surface. What remains—tannins, flavor compounds, and acids—becomes more concentrated, increasing the quality and aging potential of the wine.

ALL BARRELS SOFTEN AND ENRICH WINE'S TEXTURE
Air enters through wood's pores, exposing the wine to very slow, continuous oxidation. This causes small-scale chemical reactions that soften harsh young wines and make them feel smoother on the palate.

ONLY NEW BARRELS ADD OAK FLAVOR AND TANNIN TO WINE
Oak contains soluble flavor compounds and tannins, which are imparted to wine over time. New barrels give wine a strong, toasty flavor like that found in cognac or bourbon. However, just like a tea bag, a barrel will lose its flavor gradually with each use, approaching a neutral state by its fourth year. For most wines, 100 percent new oak is too strong, so vintners typically rotate in only 20–50 percent new barrels with each vintage.

WHAT OAK BARRELS DO TO WINE

LET WATER AND ALCOHOL OUT

LET AIR AND OXYGEN IN

TRANSFER OAK FLAVOR

OAK BARRELS VS STEEL TANKS

For the full trifecta of concentrated flavor profile, rich mouthfeel, and oaky flavor, only wood barrels will suffice. But not all wines can justify the inherent costs of patient aging in barrels.

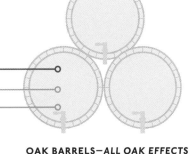

**INERT VESSELS—
NO OAK EFFECTS**
Wines that are unoaked are fermented in vessels that impart no flavor, such as stainless-steel tanks. Most white and rosé wines are unoaked, but only the lightest, youngest, and least ambitious red wines are made this way.

OAK BARRELS—ALL OAK EFFECTS
Premium red wines are fermented in steel tanks, then matured in oak barrels before bottling. Since reds are more harsh when young, they need anywhere from 3 months to 3 years to mellow in barrels, depending on the wine's style and ambition. Only the heaviest white wines see oak at all, but those that do are usually fermented in barrels from the start, as with Chardonnays. For both red and white wine, newer barrels and longer aging strengthen their oak flavors. However, older, "neutral" barrels will impart no obvious oaky taste.

> **INTENSIFIES WINE**

> **ENRICHES TEXTURE**

> **ADDS OAK FLAVOR**

OAK TREATMENTS—OAK FLAVOR ONLY
The taste of oak is appreciated by many wine drinkers, but not all wines can justify the expense of barrel aging. Instead, a shortcut is often taken for cheaper brands, where the wine is "oaked" in steel tanks using wood products like chips or staves.

WINE IN OAK OR OAK IN WINE?

A cognac-like oaky flavor is a prized hallmark of many fine wine styles, but quite time-consuming and expensive to create the old-fashioned way. Modern winemakers can use wood chips or planks to impart an oaky taste much faster and more economically. However, such oak treatments cannot replicate the other improvements that come with barrel aging, so they are rarely used other than for bargain wines.

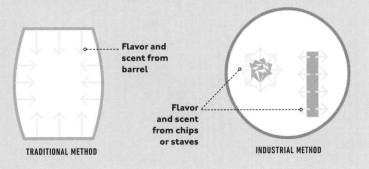

Flavor and scent from barrel

Flavor and scent from chips or staves

TRADITIONAL METHOD   INDUSTRIAL METHOD

THE TASTING

Identifying Grape Skins and Oak Barrels

COMPARE FOUR WINES AT HOME

All four of these wines are made with dark-skinned grapes in the Pinot family, but their degree of color is determined by how long the grape juice remains in contact with grape skins. Red and purple grapes can be made into white wines if their skins are removed before fermentation begins. Brief skin contact adds just a blush of color and flavor. Only the red wines receive oak-barrel treatment.

1 Taste wine 1 (white) and wine 2 (rosé). Notice how fresh they taste. This is because they are fermented chilled, with minimal skin contact, then bottled unoaked and released while young.

2 Now sample the red wines 3 and 4. Notice how fermenting warmer, with grape skins, alters their flavor. Being premium reds, both wines have been aged in oak barrels to soften and enrich their mouthfeel. But wine 4 is intense enough to benefit from the stronger flavor of newer oak, and that should be apparent on tasting.

FERMENTING WITH THE SKINS EXTRACTS NOT ONLY COLOR AND FLAVOR BUT ALSO HARSH BITTER COMPONENTS THAT SOFTEN AND MELLOW WHEN AGED IN OAK BARRELS.

WASTE NOT WANT NOT

If you need to taste with only one or two people, don't fret about waste; for a helpful tip on preserving opened wines to drink later, see Freezing Wine, p.63.

SQUEEZE IMMEDIATELY, THEN DISCARD SOLIDS

1

UNOAKED WHITE FROM RED GRAPES

PINOT GRIGIO

For example ... Italian Pinot Grigio, German Grauburgunder, American Pinot Gris, or Canadian Pinot Gris

Can you detect ...?
Low sugar/dry;
high acidity/tart;
low fruit intensity;
no oak flavor;
low alcohol/light;
no tannin

Explanation Pinot Grigio is a paler-skinned variant of Pinot Noir. While Pinot Noir looks dark purple, Pinot Grigio looks reddish pink on the vine. Since it isn't dark enough to make good red wine, its grapes are immediately squeezed, and only the clear juice is cold-fermented, making a white wine.

SQUEEZE EARLY IN FERMENTATION, THEN DISCARD SOLIDS

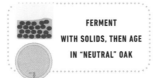

FERMENT WITH SOLIDS, THEN AGE IN "NEUTRAL" OAK

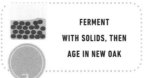

FERMENT WITH SOLIDS, THEN AGE IN NEW OAK

2

UNOAKED PINK FROM PURPLE GRAPES

3

RED AGED IN OLD BARRELS FROM PURPLE GRAPES

4

RED AGED IN NEW BARRELS FROM PURPLE GRAPES

PINOT NOIR ROSÉ

For example ... Try for an Australian one, or substitute any pink Pinot Noir from France, New Zealand, or the US

Can you detect ...?
Lowish sugar/medium-dry;
high acidity/tart;
moderate fruit intensity;
no oak flavor;
low alcohol/light;
negligible tannin

Explanation Making pink wine combines elements of both red- and white-wine making. Dark grapes are crushed and spend 6–48 hours soaking with the skins. When they have picked up just enough color and flavor, the pink juice is separated from the grape solids and cold-fermented.

MODEST FRENCH PINOT NOIR

For example ... Select a modestly priced Bourgogne Pinot Noir or another young, affordable French Burgundy, such as Mercurey or Santenay

Can you detect ...?
Very low sugar/very dry;
high acidity/tart;
moderate fruit intensity;
mild oak flavor;
medium alcohol/mid-weight;
mild tannin

Explanation Red wines extract color and flavor from the dark grape peel, along with harsh astringent tannins. Vintners typically age all but bargain reds in barrels to let them mellow before bottling. The mildest and most affordable European reds are often aged in older barrels that have largely lost their new-oak taste.

PREMIUM AMERICAN PINOT NOIR

For example ... From a California appellation such as Sonoma County or Monterey, or a substitute from Oregon, New Zealand, or Canada

Can you detect ...?
Low sugar/dry;
moderate acidity/tangy;
higher fruit intensity;
strong oak flavor;
medium alcohol/mid-weight;
moderate tannin.

Explanation Premium New World wines are typically made from riper grapes, and vintners try to maximize color and flavor extraction. Making reds more intense makes them harsher, too, so longer barrel aging is needed to soften them. Bolder grape flavors merit a stronger "seasoning" with newer barrels.

SPECIALTY STYLES: FORTIFIED WINE

The alcohol in most wines results from natural fermentation; however, a handful of styles are "fortified" with distilled spirits—usually with a crude grape brandy resembling grappa that is nearly pure alcohol. Fortified wines have more alcohol than standard wines, typically 15–20%. This makes them smell stronger and feel heavier, which is why they're served in small portions. They can be white or red, but the most popular are sweet.

HOT AND HEAVY

Heavyweight fortified wines are specialties of hot climates, since sun and heat make grapes easy to grow but also lead lighter standard wines to spoil easily.

AND NOW FOR THE HISTORY BIT

Fortified wines are historic relics that have survived because we like the way they taste. Wine casks were originally spiked with spirits by wine merchants and sea captains, not winemakers, as a preservative before shipping. Like salting fish or pickling vegetables, measures for preventing spoilage were routine for long-distance transport, especially in hot climates. This helps explain why the most popular fortified wines—from Port and Sherry, to Madeira and Marsala—were all innovations of British merchants stationed in warm-climate ports, tasked with supplying wine for a global seafaring empire. In the wineries they owned or controlled, stabilizing wine by adding alcohol was integrated into winemaking to improve quality. Strong demand for these products in Britain and the colonies led local Portuguese, Spanish, and Italian vintners to follow suit.

STYLE OUT OF NECESSITY

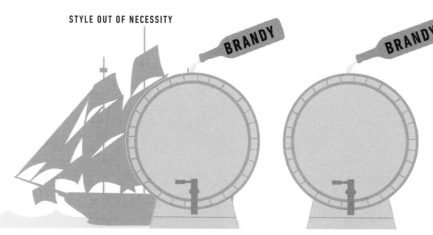

TAKE THE HISTORICAL PRACTICE
Fortify the finished wine with brandy in order to prevent spoilage during transport to distant markets.

ADAPT IT TO WINE MATURATION
Fortify the wine after fermentation but before cask-aging, to preserve it in the traditional "Sherry method."

REFINE IT TO MAKE A SWEET TREAT
Fortify mid-fermentation to kill yeasts, retain grape sugar, and preserve the wine in the "Port method."

SHERRY METHOD:

FORTIFIED AFTER FERMENTATION

Wines made by this older method are almost always white wines that are fermented to dryness first. The resulting "base wine" is then fortified with spirit and aged. Such wines may or may not be sweetened afterward, so they can range from bone-dry wines like Fino Sherry to sweet, sticky wines like Cream Sherry.

Sherry-style wines are almost always flavor-boosted to balance their strong alcohol. Wines made by the Sherry method include those below.

SPANISH SHERRY

Flavor is amplified with either oxidative aging for stronger brown Sherry or aging with special flor yeast for paler, lighter-bodied Fino Sherry (or both for Amontillado Sherry).

PORTUGUESE DRY MADEIRA

Flavor is amplified with oxidative aging at high temperature, known as maderization, for dry Sercial and Verdelho Madeiras.

ITALIAN DRY VERMOUTH

Rarely sold as wine, Vermouth is, in fact, wine that is fortified and flavored through the infusion of herbs and other botanicals.

ADDED GRAPE SUGAR (OPTIONAL): RAISIN SYRUP OR JUICE CONCENTRATE

ADDED ALCOHOL: DISTILLED SPIRIT (GRAPE BRANDY)

NATURAL ALCOHOL: PRODUCED DURING FERMENTATION

PORT METHOD:

FORTIFIED DURING FERMENTATION

This technique, also known as *mutage*, is a more recent innovation and makes only sweet wines, which may be white or red. Distilled spirit is added much earlier, during the fermentation stage. Since yeast cannot tolerate alcohol levels of more than 15%, adding brandy stops the fermentation process abruptly and guarantees a sweet dessert wine. Wines made by the Port method include those below.

PORTUGUESE PORT

Most Ports are red and come in two main styles. Russet-colored Tawny Ports are barrel-aged and nutty, while the more classic purple Ports are protected from oxidation to stay jammy and vivid.

FRENCH VINS DOUX NATURELS

White Muscats are sweeter and drunk young, but Grenache-based reds like Banyuls are drier and often given barrel aging.

SPANISH VINOS DE LICOR

Andalusian Moscatels and Pedro Ximénez from Sherry country start with sun-dried white grapes before fortification.

Other styles made this way include sweet Madeira and Moscatel from Portugal, Marsala and sweet Vermouth from Italy, and Malaga and Montilla-Moriles from Spain.

NATURAL GRAPE SUGAR: PRESERVED FROM FRESH GRAPE JUICE

ADDED ALCOHOL: DISTILLED SPIRIT (GRAPE BRANDY)

NATURAL ALCOHOL: PRODUCED DURING FERMENTATION

SPECIALTY STYLES: SPARKLING WINE

Carbon dioxide and alcohol are byproducts of fermentation, so all wines are bubbly at one stage. This fizz is usually allowed to dissipate, but some wines taste better with a little "sparkle." To capture natural carbonation, vintners tinker with the standard winemaking process, conducting the final stage of fermentation in a closed container that can trap wine's bubbles.

TRADITIONAL METHOD

The laborious "traditional method" pioneered in Champagne is still used worldwide for premium wines because of its delicious results: wines with fine, creamy bubbles, combining the refreshment of lighter wines with the opulence of richer wines. But cheaper wines are often made by less time-consuming methods—modifying the traditional method or skipping the second fermentation.

1 MAKE THE BASE WINE
A dry low-alcohol still white wine is made, often from a mix of underripe red and white grapes.

2 BOTTLE AND SWEETEN
Bottles of base wine are dosed with measured amounts of sugar and yeast, then tightly sealed.

3 SECOND FERMENTATION
Yeasts consume the sugar, generating alcohol and carbon dioxide, which gets trapped as carbonation.

4 AGE ON THE LEES
A spent yeast sediment forms after fermentation. Aging wine with these "lees" adds a doughy flavor and enriches the wine's texture.

5 CLARIFY THE WINE
After aging for between 6 months and 10 years, the clear wine is separated from its sediment by inverting bottles and controlled freezing.

6 TOP OFF AND SWEETEN
Lost volume is replaced with wine. Cane sugar is added to offset the wine's hyper-dry edge. For pink styles, red wine is used to add color.

SPARKLING WINE PRODUCTION AND CHARACTERISTICS

	CHAMPAGNE STYLE	PROSECCO STYLE	ASTI STYLE
PRODUCTION METHOD	Fermented twice, second time in sealed bottles	Fermented twice, second time in sealed tanks	Fermented once, in sealed tanks
KEY FACTOR	Flavor and texture enriched by long-term lees aging	Bottled and sold young to preserve freshness	Fermentation interrupted to retain sweetness
CARBONATION	Fine bubbles; long-lasting creamy mousse	Medium bubbles; persistent frothy mousse	Larger bubbles; short-lived foamy mousse
SWEETNESS	Most often dry to very dry	Most often dry to lightly sweet	Always sweet

TERMS OF SWEETNESS

One of the most confusing aspects of sparkling wines is the label terms that indicate sweetness, which can sound contradictory, thanks to an accident of history. When French Champagne first became popular, it was sweetened with as much sugar as modern soft drinks in the final stage, know as the "dosage." But over time, customers wanted drier styles, so vintners added less sugar and labeled these bottlings as *demi-sec* or *sec*, meaning half-dry or dry. When export markets demanded even drier wines, they had to come up with a new word to mean "drier than dry." The term *brut* was coined, meaning savage and unrefined, to convey their near-total absence of sweetening.

A REAL BRUT
Brut wines fall below the level of perceptible sweetness and dominate the realm of premium modern sparkling wines. Confusingly, "extra-dry" wines are not drier, as their name suggests, but are sweeter than the brut norm.

DEMI-SEC
FULLY SWEET

EXTRA-DRY
FAINTLY SWEET

BRUT
VERY DRY

BRUT NATURE
EXTREMELY DRY, WITH
NO ADDED SUGAR

CHAPTER CHECKLIST

Here is a recap of some of the most important points you've learned in this chapter.

- Wine is made through **fermentation**, a process in which living yeasts consume grape sugar and convert it into **alcohol and carbon dioxide**.

- Most wines are dry, especially those over 13% alcohol, because fermentation naturally **depletes grape sugar**.

- Vintners can make **sweet wine** by interrupting fermentation, by concentrating grapes before fermentation, or by sweetening a dry wine.

- White wines can be made from **grapes of any color** because the grape skins are removed before fermentation; but reds and rosés can only be made from **dark-skinned grapes**.

- White wines taste more like grape juice and are fermented cold to preserve their **fresh taste**. Red wines taste more like grape peels and are fermented warm to extract color, flavor, and tannin compounds.

- Rosés start fermentation with their **skins**, as for reds, but these are soon removed, and the wines finish fermenting as if they were whites.

- Fermenting or aging wine in oak barrels will boost its **concentration** and enrich its **mouthfeel**. If some of the barrels are new, the wine will also pick up **oaky flavors**.

- Oak aging is more necessary for red wines than whites, because it **softens and mellows** the harshness of grape-skin compounds.

- Some wines are **fortified** with added distilled spirit, raising their alcohol content to 15–20%. Most, but not all, are sweet dessert wines.

- **Carbonation** is a natural byproduct of fermentation. Most **sparkling wines** are made by re-fermenting a still wine in a sealed container to trap the bubbles.

GRAPE-GROWING CHOICES

QUALITY, INTENSITY, AND *TERROIR* Any winemaker will tell you that what happens in the vineyard is more important than what happens at the winery. Since wine is made from only grapes, every factor that affects their flavor and quality will be reflected in the glass. The land's geography—from the larger region's macroclimate to the intimate folds of its terrain— significantly shapes the flavor potential of its fruit. Farming decisions at every level have a direct impact as well: not simply when to pick the grapes or what to plant where, but the critical question of how to manage the life cycle of the land.

LOCATION, LOCATION, LOCATION

Wine's flavor is more deeply affected by the place where it is grown than most agricultural products. Everything about the vineyard informs how its wine will taste—from macro-level geographic factors like latitude to micro-level nuances like soil composition; from unchangeable features like terrain to variable conditions like the weather at harvest.

MAKING SENSE OF WINE

Many of the most confusing aspects of wine start to make more sense once you understand a few central concepts about the importance of the land.

FAMOUS NAMES

The largest wine regions, like California or Tuscany, make the cheapest everyday wines but benefit from name recognition. Premier appellations are typically tiny places that no one's ever heard of, like Rutherford or Barolo.

> A WINE'S APPELLATION, OR FORMAL REGION OF ORIGIN, IS THE MOST IMPORTANT QUALITY FACTOR LISTED ON ANY WINE'S LABEL.

Exceptional wine can only come from great vineyards. Cabernet Sauvignon may be a noble grape, but it needs very specific growing conditions: Its wine wouldn't taste good if it was grown in the Sahara or Siberia. This is why so many European wines are named for their region, like Côtes du Rhône, and not for their grapes, like Grenache.

> THE SMALLEST APPELLATIONS ARE, BY DEFINITION, THE MOST PRESTIGIOUS AND ALMOST ALWAYS MAKE SUPERIOR WINE.

Specificity about where the grapes were grown is associated with greater quality potential and therefore higher prices. There is no economic incentive to recognize a small appellation inside a larger one unless the land can make better, more distinctive wines. In Europe, the most rigorous quality standards are applied to the smallest appellations.

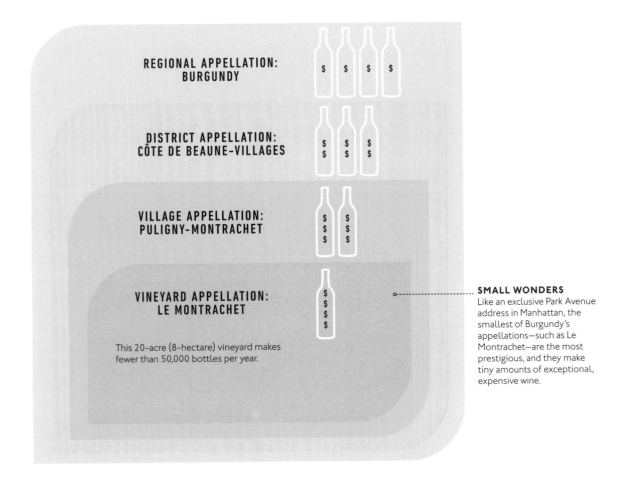

SMALLER AND SMALLER

Within any larger wine region—California, for example—the best vineyard zones establish their own smaller appellations to distinguish their wines from the pack and increase their value. Over time, these subzones—like Napa Valley—are judged by their wines' quality, and those with the best track record command higher prices. If an appellation gets famous enough, its top vintners will start the process again, carving out an even smaller subzone, as has been the case with Napa's Rutherford or Howell Mountain appellations.

REGIONAL APPELLATION:
BURGUNDY

DISTRICT APPELLATION:
CÔTE DE BEAUNE-VILLAGES

VILLAGE APPELLATION:
PULIGNY-MONTRACHET

VINEYARD APPELLATION:
LE MONTRACHET

This 20-acre (8-hectare) vineyard makes fewer than 50,000 bottles per year.

SMALL WONDERS
Like an exclusive Park Avenue address in Manhattan, the smallest of Burgundy's appellations—such as Le Montrachet—are the most prestigious, and they make tiny amounts of exceptional, expensive wine.

APPELLATIONS AND ORANGES

Appellations are formal regions of wine origin that signal the value of a vineyard's real estate. Just as oranges from Florida or Valencia command premium prices, so do wines from places like Bordeaux. But wine appellations go much further to recognize superior land, such as the legendary village of Margaux within the respected Médoc peninsula of Bordeaux. The most historic wine regions typically have the most complicated appellation structure. Burgundy's largest appellation, Bourgogne, encompasses 100 others, for example—from regions, to districts, to villages and beyond. The smallest are a few dozen *grand crus*—or single-vineyard appellations—that denote the very best Burgundy wines.

GEOGRAPHY AND CLIMATE

Grapevines need certain conditions to grow, so all wine regions have some things in common. They are in temperate latitudes with enough warm summer days to ripen grapes, but they get cold enough in winter for their vines to have a dormant season. Within these boundaries, however, there is a lot of regional variation that affects how wine will taste.

RELATIVE RIPENING

Vineyard geography is reflected in wine flavor in many ways, most significantly through its impact on ripening (see pp.68–85). Grapes ripen faster and more thoroughly in South Australia, for example, than they do in New Zealand, because they are planted closer to the equator and are not cooled on all sides by frigid waters. Coastal Tuscan vines see some clouds and rain in the growing season, while Argentina's vines grow in near-desert conditions at the base of the Andes.

GROUND CONTROL

A wine region's topography sometimes needs a boost from its geology—for example, Chardonnay grapes struggle to ripen in Chablis and need as much sun as possible in this chilly zone of northern France. Prime vineyards here require southern exposure, and even these sites need good weather to make sound wines. But the finest wines of Chablis come from its six best vineyards, known as its grands crus. They occupy a single south-facing slope and feature unusually pale, chalky soil that reflects the sun's warmth back up to the grapes from below. The combination of terrain and soil boosts ripeness dramatically, making wines from these vineyards taste unlike Chardonnays made anywhere else in the world.

MATCHING GRAPE TO REGION

Vintners juggle lots of information when deciding which grapes to plant where. Burgundy's native Pinot Noir and Chardonnay grapes, for example, are well adapted to cool growing conditions, while thicker-skinned Cabernet Sauvignon from Bordeaux requires much more warmth to ripen.

TOO HIGH: TOO DRY AND WINDY

PRIDE OF PLACE

Where grapes are grown is of great importance in wine—not just in broad terms of country or climate, but down to which side of a hill they're planted on.

SOUTHERN EXPOSURE

Pinot Noir bakes when it gets too hot, but it still needs sunlight to develop intense flavor. In Burgundy, this grape's finest wines are all grown on an escarpment called the Côte d'Or ("Golden Slope"), which faces the rising sun. For centuries, it has been recognized that the vineyards in the center of the slope make the finest wine, and many of these have been granted prestige *grand cru* status. However, grapes grown just above or just below on the hillside don't perform as well, so their vineyards are classified as either next-best *premier cru* sites or are lumped in with the more generic wines that are bottled under the name of their village. Internationally, Pinot Noir follows this model, too: Vintners in cool regions—from Canada to New Zealand—plant this grape on sunward slopes to achieve the same effect.

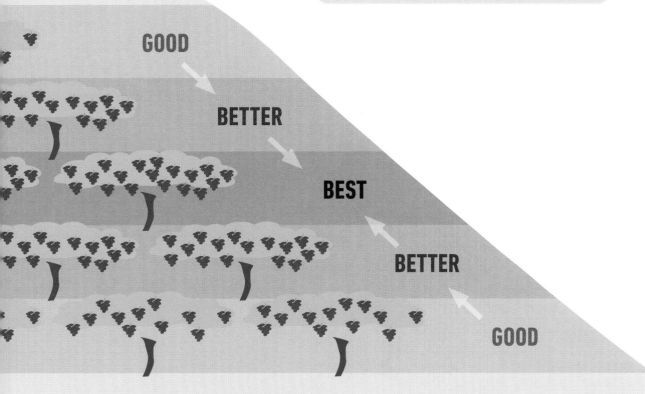

GOOD

BETTER

BEST

BETTER

GOOD

GOOD, BETTER, BEST
In Burgundy's Côte d'Or, the "best" *grand cru* sites for Pinot Noir are located mid-slope, as shown above, flanked by "good" village-level land and "better" *premier cru* sites.

TOO LOW: TOO WET AND HUMID

THE EFFECTS OF *TERROIR*

Variations in geography and climate can make wines made from the same grape variety taste remarkably different from one region to the next. But smaller variations in terrain and soil composition within the same region can also affect ripening potential, thereby altering a wine's flavor.

WHAT IS *TERROIR*?

Terroir means earth or soil in French, but the term has been adopted in wine-speak to mean location-specific flavor, roughly the "taste of the place." *Terroir* is often described as an "earthy" or "mineral" scent but can manifest itself in wine's ripeness, texture, and finish, too. Some professionals can identify a wine's vineyard by taste alone, but *terroir* is not obvious to the average wine drinker. If wine were music, *terroir* wouldn't be a song's melody or arrangement as much as the distinctive acoustics of a specific venue or recording studio—you'd have to be an enthusiast to recognize it in tracks by different artists.

Reading about *terroir*, it's natural to wonder if there's actual dirt in your wine. There's not, but it's been known for centuries that vineyard soil plays a strong role in wine flavor, and it has become clear more recently that farming choices can amplify or suppress wine's *terroir*. The mechanisms aren't fully understood, but the interplay of life cycles in the vine's environment appear to hold the answer, particularly microbiological activity such as soil renewal and fermentation. Certainly, systemic chemical treatments like herbicides obscure *terroir* traits in wine, suggesting a tasteable connection with the web of life belowground that leads many premium vintners to practice organic or natural farming.

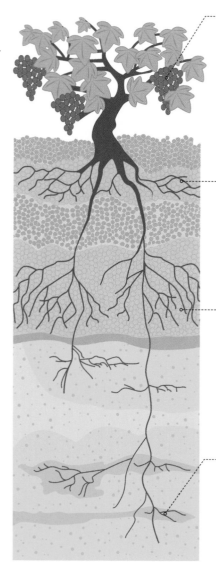

FINGERPRINT OF FLAVOR
Vineyard-specific nuances of taste that might not be apparent in fresh grapes can be exposed in wine by fermentation.

SOIL FAUNA
From earthworms to fungi, a complex web of life sustains healthy vineyard soils and contributes to wine's distinctive *terroir*.

SOIL COMPOSITION
Winemakers have known for centuries that variations in mineral and nutrient content result in consistent flavor changes in wine.

SOIL DRAINAGE
Vines that must dig deep for water are the most resilient, producing better fruit in both droughts and stormy seasons.

TASTEABLE ENVIRONMENT
Vineyard soils and ecosystems are major quality factors in grape growing, and their effects can be tasted in the wine glass.

TIME TO GET FUNKY

We know microbiology can create great flavor complexity, as with raw-milk cheeses like Roquefort or Gruyère. Wine- and cheese-making traditions have led the Old World to embrace a little funky individuality, and earthy *terroir* traits are often celebrated in European wines. New Worlders tend to view microbes of all kinds as threats to be eradicated, so their wines more often emphasize squeaky-clean fruit flavors, especially for value wines.

THE IMPORTANCE OF SOIL

The concept of *terroir*—that we can taste the effects of vineyard environment—is a powerful tool for understanding how the fine-wine world works. Complex appellation systems and vineyard-ranking hierarchies are all attempts to organize wines around their *terroir*.

One clear example of the importance of soil is found in Bordeaux. Cabernet Sauvignon makes the region's top wines, but its thick skins need ample heat to ripen fully, so this variety performs much better in the well-drained gravels that line the river's "Left Bank" than in the moist clay found everywhere else. It's not a coincidence that in a ranking of top red wine estates for the 1855 World's Fair, later adopted as Bordeaux's *grand cru* classification, 90 percent were clustered in four adjacent gravelly villages on the Médoc peninsula.

Early-ripening Merlot makes more reliable wines on clay. However, because the land is more marginal and the wine less prized, most is sold as everyday regional Bordeaux wines.

MANY MEANINGS

Narrowly applied, a wine's *terroir* is the individuality of taste and smell that its vineyard imparts. More broadly, a region's *terroir* is its unique combination of flavor-impacting factors, including climate, landscape, and soils.

MONEY IN THE BANK
The deep, dry gravel beds of Bordeaux's Médoc zone, known as the "Left Bank," boost ripening potential enough to favor Cabernet Sauvignon, which makes most of the region's most expensive wines. The damp clay that dominates the region's "Right Bank" slows down ripening and is therefore better suited to growing Merlot.

Left Bank: warmer gravel

Right Bank: cooler clay

FARMING FOR QUANTITY OR QUALITY?

Even if they were grown in the same appellation, grapes for easy-drinking value wines would be cultivated very differently from those destined for super-premium luxury wine. Artisanal wines taste different from mass-produced wines for the same reasons that heirloom tomatoes from your garden taste different from supermarket tomatoes.

BULK AND PREMIUM WINES

Vineyards for bulk wine and premium wines are managed with opposing priorities in terms of productivity and agricultural philosophy. Value-oriented vintners purchase their grapes from growers who operate at maximum efficiency to stay competitive. Farming for the mass market may require fertilizing, irrigating, and mechanized spraying to boost crop volume and keep prices low. Ambitious vintners, however, farm by hand, often growing grapes on their own land. Since bumper crops dilute flavor, they prune vines rigorously to deliberately suppress the number of grape clusters per vine. To improve wine quality, they will also manage their vineyard with little or no chemical intervention and will often plant on steeper, hardscrabble sites that are less fertile. Such vineyards are more difficult to cultivate and yield far less fruit, but their grapes make better, more age-worthy wines.

MAXIMIZING VINEYARD YIELDS
Vines can produce bumper crops of grapes with adequate sugar for winemaking when yield is maximized, making for affordable wine. However, doing so weakens their flavor and can result in unbalanced wine in terms of components like acidity and alcohol.

FARMING FOR QUANTITY

INCREASES CROP VOLUME
TO MAXIMIZE EFFICIENCY

CONVENTIONAL FARMING THAT
EMBRACES CHEMICAL TREATMENTS

USES MECHANICAL CULTIVATION

WHO GREW THE FRUIT?

Most wineries do not own or manage vineyards. Instead, they purchase fruit from grape growers who own and cultivate vineyards. However, a minority of vintners are what is known in the trade as "estate" wine producers, meaning that they do own and manage their own vineyards. Vintners have far more control over critical quality factors like crop yield when they grow their own grapes, which helps explain why almost all of the world's best wines are estate wines. Consequently, claims of vineyard ownership on the label suggest superior wine quality.

ESTATE OR NOT ESTATE?

Wines made entirely from vineyards owned or controlled by the winery are called "estate wines." Most wine-producing countries provide some means to convey this information on the label and regulate its use tightly. New World countries use phrases like "estate bottled" or "estate grown," and permit them only for those wines grown entirely on property that is owned or managed by the winery. The presence of the word "estate" in a winery's name does not guarantee that every wine they make is an estate wine though, just that they own at least some vineyard land.

In Europe, discerning ownership can be harder because each region has its own terms for a wine estate. In Burgundy, the honorific term *domaine* is reserved for estate wines. This is proudly included in the winery names of those who produce only estate wines to distinguish them from so-called *négociant* wines made from purchased fruit. The terms *château* in Bordeaux or *tenuta* in Tuscany have similar meanings. When in doubt about whether a European winery grows its own grapes, consult the fine print. If it doesn't reference land ownership or agriculture, the wine is likely to have been made from purchased fruit.

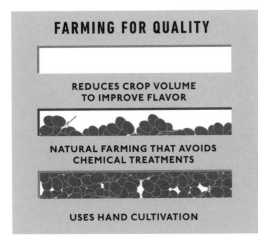

FARMING FOR QUALITY

REDUCES CROP VOLUME
TO IMPROVE FLAVOR

NATURAL FARMING THAT AVOIDS
CHEMICAL TREATMENTS

USES HAND CULTIVATION

RESTRICTING VINEYARD YIELDS
Top vintners prune their vines back to limit crop volume because grapevines do a more thorough job of ripening grapes when they bear fewer of them. Improving wine quality requires sacrificing crop volume, which helps explain why prices of premium wines can climb so high.

ORGANIC GRAPES AND NATURAL WINES

Prior to the Industrial Revolution, all farming was organic by today's definition. Nowadays, the agricultural norm involves boosting productivity with fertilizers and fungicides, herbicides and pesticides, all of which are routine for grape-growing, too. However, vintners have long known that reducing or eliminating chemical use in both the vineyard and the winery increases wine quality.

ORGANIC FARMING REQUIREMENTS

- Only natural inputs permitted—no synthetic substances used
- No genetic engineering
- Verified by inspections and certification
- Strict labeling standards

BETTER ALL AROUND
Organic farming is not simply about doing what's best for the environment. It's also about farming for quality—growing the best possible fruit to make the best possible wine.

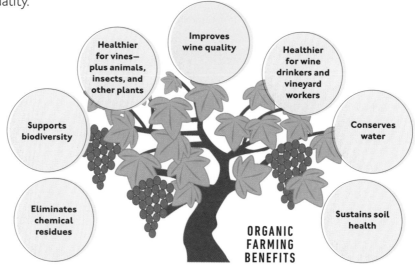

Improves wine quality

Healthier for vines— plus animals, insects, and other plants

Healthier for wine drinkers and vineyard workers

Supports biodiversity

Conserves water

Eliminates chemical residues

Sustains soil health

ORGANIC FARMING BENEFITS

ORGANIC AND BIODYNAMIC

Not all premium wines are made from organic grapes. However, the finer the wine, the more likely it is to have been grown as naturally as possible, since this approach improves quality. Some vintners make an extra effort by certifying their grapes or wines as organic, meaning they were grown without the use of synthetic inputs or genetically modified organisms. A few go further by adopting even more exacting standards of natural farming to qualify their fruit as biodynamic. Biodynamic farming embraces the preindustrial traditions of European agriculture and treats the vineyard as a self-contained ecosystem.

Where organic certification simply prohibits use of human-made materials, biodynamic certification requires specific actions be taken, often synchronized with solar, lunar, and astral cycles. Some people may scoff that biodynamic practices seem more like superstition than science, but it's hard to argue with results. Wines produced this way consistently display more individuality, more *terroir* traits, and a longer finish—all qualities prized by collectors of fine wine. However, natural farming is quite expensive. Organic wines tend to cost more than conventionally farmed wines, while biodynamic wines are most often expensive luxury wines.

MUCH ADO ABOUT SULFUR

Most countries agree on what counts as organic farming, but each has its own standards for organic winemaking. As a result, the meaning of "organic" on a wine label is not straightforward and does not always reference farming alone. For example, chemical additives are prohibited in both organic farming and winemaking. However, adding a smidgen of sulfur dioxide to grapes before fermentation, originally by flame purification of fermentation vessels, has been essential for hygienic winemaking for millennia.

Omitting this step does not appreciably change a wine's sulfur content once it is bottled, since sulfur dioxide is a natural byproduct of fermentation and, when added to grapes, largely settles out as sediment. Choosing not to add sulfur at this stage does destabilize wine though, allowing microbiological activity and oxidation to shorten its shelf life.

Adding more sulfur to finished wine after fermentation as a preservative is a practice that is always prohibited for organic wines, and capped for all wines, as explained below. However, rules on added sulfur before fermentation vary widely. Most countries permit limited levels for their organic wines, in line with fine-wine norms. Some have stricter rules, most notably the US, which prohibits sulfur use entirely for organic wines. As a result, many American wines are labeled as "made from organic grapes" rather than as "organic wine."

WHAT IS NATURAL WINE?

Claims of "natural winemaking" have become fashionable, but legal definitions are scarce. In theory, natural wines are those made with no "unnatural" interventions, from vineyard to winery to glass, but what counts as unnatural is open to interpretation. Wines marketed as "natural" are often revivals of ancient styles, as with "orange" wines made by fermenting green grapes on their skins in oxidative conditions, or unfiltered and semi-sparkling "pét-nat" wines. Many can be delicious, with a vibrancy reminiscent of raw milk cheeses. Some are lightly sulfured, others are not, but all are controversial. Fans love their individuality and idealism, while detractors find them unclean and prone to premature spoilage.

HIGH SULFUR DIOXIDE: 100–400 PARTS PER MILLION

Wines that are sweet and/or low in natural preservatives like tannin have the most sulfur added, usually both before and after fermentation. This includes dessert wines, white wines, and rosés, as well as mass-produced bulk wines and bargain wines.

LOW SULFUR DIOXIDE: 30–100 PARTS PER MILLION

Less sulfur is needed for wines that are dry, red, or high quality, so less sulfur is added and typically only before fermentation. Most organic wines are made this way, too, as are biodynamic wines and those labeled "made with organic grapes."

NO ADDED SULFUR DIOXIDE: <30 PARTS PER MILLION

Sulfur dioxide is naturally present in both grapes and wine, so all wines contain some, even if none is added, as with organic wines from the US and "no sulfur added" natural wines. Sulfites can be stripped out but at the expense of flavor.

THE TASTING

Identifying Vineyard Factors

COMPARE WINES FROM DIFFERENT APPELLATIONS AND QUALITY LEVELS AT HOME

Sample these four wines side by side to get a sense of how vineyard location and farming techniques affect wine style.

1 Bear in mind that both pairs of wines are made with the same grape variety and hail from the same wine region. However, in each case the first wine is a more generic, affordable style from a bigger appellation, while the second is a more ambitious premium wine from a smaller sub-appellation.

2 Notice how, although the pairs share a basic resemblance, wines 2 and 4 are riper and more intense. You should also be able to detect a more distinctive aromatic profile. Affordable wines from large "generic" appellations tend to be milder in flavor, shorter in finish, and less distinctive than premium wines from smaller, more prestigious appellations.

DIFFERENCES IN GEOGRAPHY AND FARMING PRACTICES CAN PRODUCE WINES THAT TASTE DIFFERENT EVEN WHEN THE SAME GRAPE IS USED.

1 MODEST WHITE BURGUNDY

For example ... French Mâcon Blanc or a similar modest white Burgundy, like Mâcon-Villages or Bourgogne Blanc

Can you detect ...?
Low sugar/dry;
high acidity/tart;
low fruit intensity;
no oak flavor;
medium alcohol/
mid-weight;
simple, easygoing,
refreshing

Explanation Wines bottled under large appellations—like this regional Burgundy—can be perfectly delicious and refreshing. However, their vineyards are more serviceable than exceptional, and the wines may be blended from anywhere in the region.

2 PREMIUM WHITE BURGUNDY

For example ... French Pouilly-Fuissé or a similar premium white Burgundy, like Meursault or Chassagne-Montrachet

Can you detect ...?
Low sugar/dry;
medium acidity/tangy;
medium fruit intensity;
mild oak flavor;
higher alcohol/heavier;
more rich, intense,
flavorful

Explanation Wines bottled under smaller sub-appellations—like this village-level Burgundy—have more quality potential, and most will be noticeably more intense. This is especially true in Europe, where quality regulations are stricter in smaller appellations.

WASTE NOT WANT NOT

If you need to taste with only one or two people, don't fret about waste; for a helpful tip on preserving opened wines to drink later, see Freezing Wine, p.63.

3 MODEST AMERICAN PINOT NOIR

4 PREMIUM AMERICAN PINOT NOIR

For example ... California Pinot Noir; or alternatively a bargain Shiraz from Southeast Australia

Can you detect ...?
Paler color;
low sugar/dry;
moderate acidity/tangy;
medium fruit intensity;
mild oak flavor;
medium alcohol/
mid-weight;
young, simple, cheerful

Explanation Farming may be less regulated in the New World, but wines from huge appellations like California still tend to be blends sourced from regions of little pedigree. The wines can be enjoyable but are usually simple crowd-pleasers, without much depth or complexity.

For example ... Carneros Pinot Noir or one from Russian River or Sonoma Coast; or alternatively a premium Australian Barossa Shiraz

Can you detect ...?
Darker color;
low sugar/dry;
moderate acidity/tangy;
higher fruit intensity;
stronger oak flavor;
higher alcohol/heavier;
more dense, aromatic,
and complex

Explanation Where conditions are ideal for making better wines, sub-appellations are formed. These vineyards naturally produce deeper, more characterful wines, and by commanding higher prices, the vintners can justify lowering yields and improving quality.

CHAPTER CHECKLIST

Here is a recap of some of the most important points you've learned in this chapter.

- The most important **quality factor** on a wine's label is its **appellation**, or formal region of origin.
- Smaller appellations almost always make **better wine** than bigger appellations.
- In large wine regions, like Burgundy or California, the best vineyard zones establish their own **sub-appellations** to distinguish their wines.
- **Vineyard geography**, including a region's climate and terrain, has a marked impact on **ripening**.
- It is common sense that grapes are rarely planted where they don't **perform well**.
- Geographic and climatic variations mean wines from the same grape variety **taste very different** from one region to the next.
- The French term **terroir** is used in wine-speak to mean location-specific flavor, roughly translating as the **"taste of the place."**
- Vineyards growing grapes for **bulk wines** and those growing grapes for **premium wines** are managed with opposing sets of priorities in terms of their **productivity** and **agricultural philosophy**.
- The finer the wine, the more likely it is that its grapes were cultivated as **naturally as possible**, with minimal chemical treatment. They also tend to have a longer finish and more **individuality of flavor**.
- Large **"generic" appellations** tend to produce wines that are milder in flavor, shorter in finish, and less distinctive than premium wines. They are also more **affordable**.

CULTURAL
PRIORITIES

HISTORY, TRADITION, AND INNOVATION Wines from different countries taste different, but not simply because of geography. Humankind plays as great a role as nature in determining the outcome of winemaking, steering the process toward desired results that vary from one culture to the next. The history of wine's development in Europe and eventual expansion during the colonial era helps explain many of wine's mysteries. Globalization may be in full swing, but there remains a tasteable distinction between most Old World and New World wines—a reflection of differences in heritage and gastronomic sensibilities.

OLD WORLD OR NEW WORLD?

One of the most useful tools that wine professionals rely on is the distinction between wines made in Europe, or the Old World, and those made in the Americas and southern hemisphere, known collectively as the New World.

DIFFERENCES IN STYLE

Terms like "Old World" and "New World" may sound archaic, but they are still used in the wine trade because they are still relevant. Not only are European wines often labeled according to their own unique rules (see pp.50–51), but they typically taste different from other wines in ways that can be anticipated before you pull a cork. Even when the same grape variety and same winemaking methods are used, Old World and New World wines generally follow a tasteable pattern, as described below.

OLD WORLD FLAVOR PROFILE
• More "traditional"—designed to taste best with food;
• Lighter in weight/alcohol;
• Lower in sweetness (among dry styles);
• Higher in acidity;
• More subtle and earthier in fruit character;
• Milder in oak flavor (where oak is used);
• Harsher in tannins (reds only).

NEW WORLD FLAVOR PROFILE
• More "modern"—designed to taste best alone;
• Heavier in weight/alcohol;
• Higher in sweetness (among dry styles);
• Lower in acidity;
• Bolder and jammier in fruit character;
• Stronger in oak flavor (where oak is used);
• Softer in tannins (reds only).

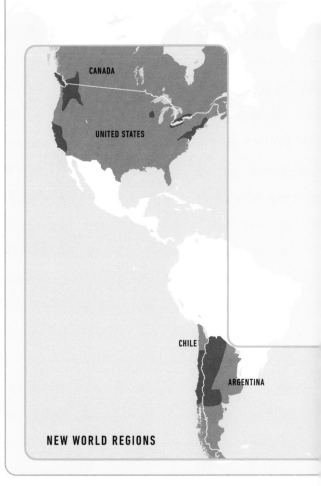

THE WORLD OF WINE

CANADA

UNITED STATES

CHILE

ARGENTINA

NEW WORLD REGIONS

NON-GRAPE VARIABLES

Such observable differences as those listed opposite do not derive from the type of grapes used in winemaking, since all fine-wine varieties are of European origin. Instead, they result from differences in two other variables that affect how wine tastes: the vineyard environment and human winemaking decisions.

THE GEOGRAPHY AND CLIMATE OF THE PLACE

Old World wine regions are almost always cooler and cloudier than the New World wine regions that grow the same grape varieties. As a result, European grapes tend to achieve lower degrees of overall ripeness.

Most New World regions are considerably sunnier, warmer, and drier in climate, which makes it easier to achieve higher degrees of grape ripeness.

THE HISTORY AND CULTURE OF THE PEOPLE

Influenced by centuries of winemaking heritage, Old World wines are often geared toward time-honored goals like aging gracefully and flattering local cuisine. New World wines tend to be shaped by a spirit of innovation and more reliant on technology. They are typically designed around very different priorities, such as pleasing a global audience on first sip and impressing wine critics.

KEY

---- MAIN WINE-PRODUCING COUNTRIES

—— MAIN WINE-GROWING AREAS

OLD WORLD REGIONS

GERMANY

FRANCE

AUSTRIA

SPAIN

ITALY

PORTUGAL

GREECE

SOUTH AFRICA

AUSTRALIA

NEW ZEALAND

WINE HISTORY IN EUROPE

Wine as we know it developed in Europe, as did all of the grape varieties used around the world today. Many puzzling aspects of today's European wine make more sense when we consider their historical context.

NORTH VS SOUTH

Southern regions on the Mediterranean, like Italy and Spain, have the longest winemaking history, the greatest proliferation of native grape varieties, and still make the most wine in terms of volume. However, Europe's best wines, in quality terms, historically came from cooler regions farther north, such as France and Germany, where a few specific grapes made smaller amounts of finer wine. This pattern was shaped by the confluence of two factors in the Middle Ages—one environmental and one socioeconomic:

- Wine quality improves when challenging vineyard conditions, such as cold climates or stony slopes, reduce total fruit production per vine (see pp.156–157).
- Investing time, land, and effort into improving wine quality might not have been worthwhile for the average medieval farmer, but made sense for the powerful Christian monasteries of central France.

BEFORE TECHNOLOGY

European winemaking traditions were well established long before the industrial and technological revolutions. Farming was always organic by definition, and winemaking was based on generations of trial and error, not on a mastery of fermentation's complex chemistry. Before wines could be refrigerated or stabilized, they needed to be as dry and age-worthy as possible. And since wine was rarely served without food, vintners naturally favored styles that flattered their local cuisine, even if they seemed a little sour or bitter on first sip.

THE OLD WORLD TIMELINE

c. 8000 BCE

Current evidence suggests that pure grape wines were first made in the Caucasus area of modern-day Georgia, the native region of the *Vitis vinifera* vine species. The practice migrated from the Black Sea into the Mediterranean basin in the Stone Age, and ancient civilizations—from the Phoenicians to the Greeks—spread winemaking around its shores, planting mostly in places where olives thrived. Vines were easiest to cultivate and most productive in these regions—zones with mild winters and dry, sunny summers. Wine soon became a staple of the Mediterranean diet alongside olive oil.

c. 100 BCE TO c. 200 CE

Vines were first planted in cooler zones when the Romans pushed northward in the mid-1st century BCE. Where olive trees give way to oak forests, cultivating wine grapes becomes more challenging: with less summer warmth, fruit does not ripen as easily. Only select varieties could adapt to chilly conditions and made the best wine when planted in the sunniest sites. Cooler northern vineyards often produced less fruit per vine than those in warmer zones farther south, but over time their wines proved to be more concentrated and more resistant to spoilage.

HISTORY OF QUANTITY VS QUALITY

Prior to the Roman era, wine was only made in southern Europe, where vines thrive easily and are prolific. Fine wines later emerged from colder regions, nearer the vine's northern limits, since grapes that struggle to survive yield less fruit and make better wine.

KEY

— PRE-ROMAN VINEYARD AREA

— LIMITS OF GRAPEVINE RANGE

★ IMPORTANT WINE REGION

5TH TO 11TH CENTURIES CE

Most of those making wine had little incentive to work harder to produce smaller amounts of wine in riskier weather conditions. Choosing quality over quantity might never have caught on outside a few enclaves of the ruling class if Europe had not been Christianized in the Middle Ages. With wine symbolizing the blood of Christ in the rite of communion, it had an exalted status. Some of the most influential monastic orders of the medieval era were based in France's Burgundy region. These ascetic monks helped transform wine from a rustic dietary staple to an elegant luxury product.

12TH TO 15TH CENTURIES

Modern grape growing and winemaking methods largely derive from those practiced by Cistercian and Benedictine monasteries in the medieval kingdom of Burgundy. Their expertise endured in part because they recorded their successes and failures. As these influential orders expanded, they spread their wine practices that valued wine quality over quantity. These flourished most in places where vines could survive but were less prone to Mediterranean-style bumper crops. By the end of the Middle Ages, regions like Burgundy and Bordeaux were famed for their exceptional wine.

THE FINE-WINE ICONS OF FRANCE

The French may not have been the first to make wine, but they were the first to pursue quality systematically on a grand scale. France had a head start on fine wine by a few hundred years and, outside of German rivalry in white wines, faced little competition before the 1800s. As a result, for more than five centuries, anyone anywhere who wanted to improve the quality of their wine would naturally look to France.

GROUND ZERO FOR FINE WINE

Today, France has many rivals in making exceptional wine, both within the EU and abroad. However, its early dominance in fine winemaking resulted in international adoption of French grapes, French styles, and French methods.

GLOBAL STANDARD-BEARERS

Outside of Europe, almost all fine wines are made with grapes that are native to France in the image of famous French wines like white Burgundy, red Bordeaux, and sparkling Champagne. Even in countries that have their own wealth of native grape varieties, such as Italy, Spain, Portugal, and Greece, Europe's wines are generally made using French techniques, and most are aged in French oak barrels. Some even supplement local grapes like Tuscan Sangiovese or Castilian Tempranillo with famous French grapes, like Cabernet Sauvignon, to improve their appeal. And when winemaking rules were standardized throughout the European Union (EU), the quality-oriented French system was adopted for all: naming wines for their place of origin, regulating permitted grapes and maximum yields, and establishing formal hierarchies based on wine quality.

Six of France's wine zones are far more influential than the rest, and three of these have a dominant presence in the fine-wine world. Burgundy, Bordeaux, and Champagne are well worth learning a little about because of their global significance as the archetypes for the majority of fine wines. All Chardonnay and Pinot Noir is made in the image of French Burgundy. Every Cabernet Sauvignon and Merlot wine is modeled on French Bordeaux. Every wine with bubbles is, in some way, paying homage to French Champagne.

The wines of three more regions—the Rhône Valley, the Loire Valley, and Alsace—may be less well known to wine drinkers, but they loom large for international vintners as the archetypes for wines like Shiraz, Sauvignon Blanc, and Pinot Grigio.

THE BIG SIX FRENCH WINE REGIONS

While many regions of France make wine, six are of disproportionate importance in the global wine marketplace: primarily Burgundy, Bordeaux, and Champagne, but also the Rhône Valley, the Loire Valley, and Alsace.

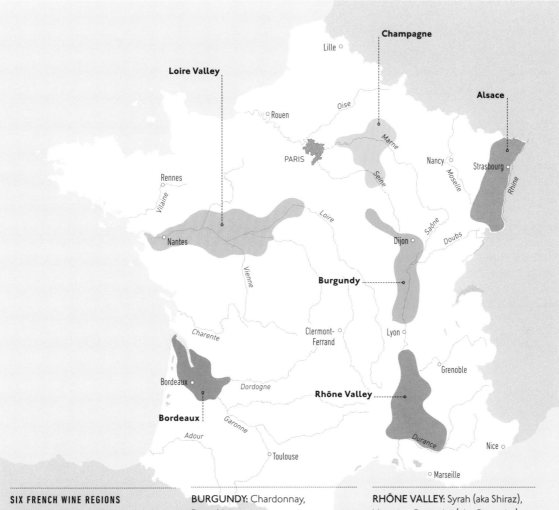

SIX FRENCH WINE REGIONS AND THEIR ICONIC STYLES

A significant proportion of the world's finest wines are inspired by a short list of French wines, whose regions and grapes are named here.

BURGUNDY: Chardonnay, Pinot Noir

BORDEAUX: Cabernet Sauvignon, Merlot, Malbec, Sauvignon Blanc

CHAMPAGNE: sparkling Chardonnay/Pinot Noir blends

RHÔNE VALLEY: Syrah (aka Shiraz), Viognier, Grenache (aka Garnacha) blends

ALSACE: Pinot Gris (aka Pinot Grigio), Riesling, Pinot Blanc

LOIRE VALLEY: Sauvignon Blanc, Chenin Blanc, Cabernet Franc

WINEMAKING IN THE NEW WORLD

Winemaking arrived in the Americas and southern hemisphere during their colonization by Europeans, and these regions remain relative newcomers, especially with regard to fine wines. Only since the 1960s or '70s have New World wines seriously rivaled Old World wines in quality—but in doing so they sparked a global wine boom and revolutionary changes in how wine is made and sold worldwide.

MORE MODERN, LESS TRADITIONAL

As a general rule, wines from New World regions share "modern" sensibilities that distinguish them from the more "traditional" wines of Europe.

NECESSITY AND INVENTION

Early New World vintners were pioneers in new lands. With no local winemaking heritage, the natural choice was to follow European "recipes." However, because their circumstances were so radically different—in terms of climate, technology, and markets for their products—their wines tasted radically different, too. Where the Old World classics tend to be lighter and milder, leaner and more earthy, New World wines tend to be bolder and heavier, more ripe and fruity in scent. This happens because most New World wines are grown in warmer regions and designed for instant gratification, while wines from cooler Old World regions are customarily more food-oriented and built to last.

THE NEW WORLD TIMELINE

16TH TO 18TH CENTURIES

Productive workhorse grape varieties were introduced to most New World colonies from Europe early on. In some regions, initial experiments were unsuccessful, as in North America, where vinifera vines lacked resistance to a local insect pest. Where vines did thrive, most wines remained rudimentary for quite some time. But with growing prosperity, ambitious vintners tried their hand at more famous grapes from French regions like Bordeaux and Burgundy to improve their wines. Some early experiments were recognized quickly for their quality potential, as with the sweet wines of South Africa's Constantia region.

19TH TO MID-20TH CENTURIES

Most New World vineyards were planted in warm, sunny regions like California and South Australia, where over-ripeness and drought were greater dangers than under-ripeness and rain. Winemakers had to improvise and, without local traditions to lean on, turned to science and technology for guidance. Most wines were simple everyday styles, and the most urgent commercial needs were to improve efficiency and increase production. Innovations in irrigation, mechanization, and chemical farming were pioneered and widely embraced. However, fine wines from quality-oriented estates also improved by leaps and bounds in this period.

NEW WORLD EXPLORATION

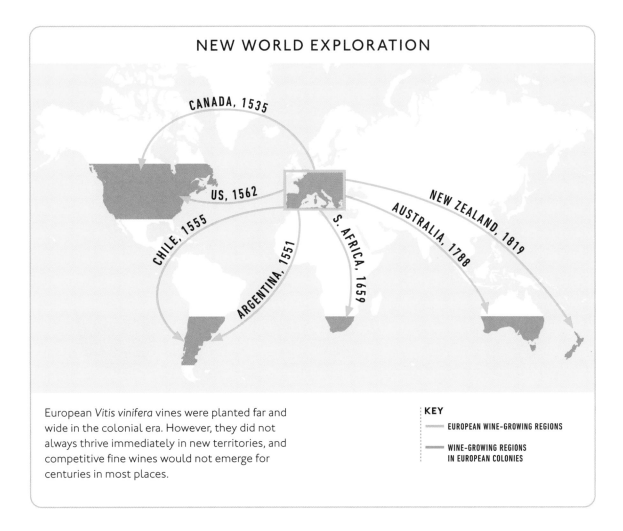

CANADA, 1535

US, 1562

CHILE, 1555

ARGENTINA, 1551

S. AFRICA, 1659

NEW ZEALAND, 1819

AUSTRALIA, 1788

European *Vitis vinifera* vines were planted far and wide in the colonial era. However, they did not always thrive immediately in new territories, and competitive fine wines would not emerge for centuries in most places.

KEY

— EUROPEAN WINE-GROWING REGIONS

— WINE-GROWING REGIONS IN EUROPEAN COLONIES

LATE 20TH CENTURY

New World wines improved rapidly in the postwar period, and by the 1980s the best could rival their French archetypes in quality. However, they didn't taste identical to Old World wines; they had their own fruitier flavor profile that was often stronger and bolder due to greater ripeness. This resulted from differences in geography, of course, but also from the differences between traditional winemaking practices in Europe and more modern techniques used in this part of the world. Also, lacking traditional pedigrees, New World wines needed to wow on first sip to compete, not just work well with food.

EARLY 21ST CENTURY

Today's competitive global market is still changing how wines taste. The once stark stylistic divide between Old World and New World wines is becoming less clear-cut over time. European vintners are making wines that are riper and more ready to drink to compete with the upstarts from the New World. Winemakers in the Americas and southern hemisphere are producing wines that are more refreshing and flattering to food as they hit their stride. Ultimately, though, you can still expect more emphasis on instant gratification from the New World and more restrained food-oriented sensibility from the Old World.

THE TASTING

Identifying Old World and New World Styles

COMPARE EUROPEAN AND INTERNATIONAL WINES AT HOME

Sample these wines side by side to get a sense of how variations in geography and culture combine to produce wines that taste very different even when the same grape is used.

KEY DIFFERENCES

New World wines tend to be more modern: riper and fruitier, designed for a strong first impression. European Old World wines tend to be more traditional: leaner and drier, designed to be flattering food partners.

OLD WORLD WHITE WINE

NEW WORLD WHITE WINE

OLD WORLD RED WINE

For example ... French Loire Valley Sauvignon Blanc, such as Sancerre, Pouilly-Fumé, Touraine, Quincy; or substitute with unoaked white Bordeaux

Can you detect ...?
Very low sugar/very dry;
very high acidity/very tart;
medium-low fruit intensity;
no oak flavor;
low alcohol/light;
noticeable "earthy" scent, like wet leaves and stones

The French way Sauvignon Blanc barely ripens in northern France, so these wines are austere and food-oriented.

For example ... New Zealand Sauvignon Blanc; or substitute with Chilean Sauvignon Blanc or an American version from California or Washington

Can you detect ...?
Low sugar/dry;
high acidity/tart;
medium-high fruit intensity;
no oak flavor;
medium alcohol/mid-weight;
very "fruity" scent, like fresh tropical fruit

The New Zealand way This island nation also makes cool-climate, unoaked Sauvignon Blanc but in a much more modern style that is riper and easier to enjoy without food.

For example ... Italian Primitivo (aka Zinfandel) from southern Italy's Puglia region; or alternatively, a French Bourgogne/ Pinot Noir

Can you detect ...?
Very low sugar/very dry;
very high acidity/very tart;
medium fruit intensity;
mild oak flavor;
medium alcohol/mid-weight;
noticeable "earthy" scent, like dried leaves or root vegetables

The southern Italian way Southern Italian vintners prioritize seafood-friendliness in their wines, so grapes are harvested early enough to retain vibrant acidity.

LEAN VS LUSH

Both pairs of wines are made with the same grape variety and in a similar way. But in each case, the European one will taste more "traditional" in style: lighter and leaner, lower in sugar and higher in acidity, and with aromas that seem more earthy and less fruity. The New World wine will be heavier and riper, slightly higher in sugar, and lower in acidity. It will feature more dessert-like smells, even if it is dry, and is more likely to have noticeable oak flavor.

NEW WORLD
RED WINE

For example ... Red Zinfandel (aka Primitivo) from any region of California; or alternatively, a California Pinot Noir

Can you detect ...?
Lowish sugar/dry or faintly sweet; moderate acidity/tangy; high fruit intensity; stronger oak flavor; high alcohol/heavy; very "fruity" scent, like jam or prunes

The American way Known as Zinfandel in California, the same grape gets different treatment in the US, where wines are judged on first sip and more often served with red meats and sweet sauces.

CHAPTER CHECKLIST
Here is a recap of some of the most important points you've learned in this chapter.

- Even when using the same **grape variety** and same **winemaking methods**, Old World and New World wines often taste different.

- In Europe (the Old World), grapes are generally grown in **cooler regions**, so they tend to achieve lower degrees of **overall ripeness**.

- New World regions are often considerably **sunnier, warmer, and drier**, making it easier to achieve higher degrees of grape ripeness.

- Wines made in the Old World are often geared toward time-honored ideals such as **aging gracefully** and **flattering food**.

- Shaped by **innovation** and more reliant on **technology**, New World wines are typically built around pleasing a **global audience** and impressing **wine critics** on first sip.

- Southern regions on the Mediterranean have the longest **winemaking history**. They also have the greatest number of **native grape varieties** and make the most wine.

- The French were the first **to pursue wine quality systematically** on a grand scale and, for more than 500 years, provided the model for anyone anywhere who wanted to **improve the quality** of their own wine.

- Six of **France's wine zones** are far more **influential** than the rest: Burgundy, Bordeaux, Champagne, the Rhône Valley, the Loire Valley, and Alsace.

- Winemaking arrived in the **Americas and southern hemisphere** when those areas were **colonized by Europeans**. These regions are relative **newcomers** to the wine world, especially in terms of fine wine.

DISCOVERING
WINE
GRAPES
----- AND -----
REGIONS

ONCE WE ARE COMFORTABLE with the big picture, it's time to get acquainted with the major players in the wine world. The grape varieties used to make wine and the regions where wine is made are the most important variables in how a wine will taste. Now that we understand more about how grape flavors shift with ripeness and how geography and culture can affect winemaking outcomes, studying grapes and regions will be much more productive. There are dozens of wine grapes and hundreds of wine appellations, and it isn't necessary or even useful to study them all. But there is a short list of top grapes worth remembering because they are so popular and so influential. And every wine lover should familiarize themselves with the world's main wine regions. After all, just a few dozen special places on earth are blessed with the right conditions to make great wine, and you'll want to explore them all.

MUST-KNOW
WINE
GRAPES

THE TOP 10 VINE VARIETIES There are thousands of grape varieties, but the vast majority of modern wines are made with only a few dozen. Of these, a handful have earned celebrity status for their quality and popularity with wine drinkers, such as Chardonnay and Riesling, Cabernet Sauvignon and Pinot Noir. Fine wine as we know it was pioneered in France, so most of these rock star grapes are of French heritage, but they are now truly international, cultivated far and wide. Since so many modern wines are labeled by grape variety, it's well worth getting to know the members of the global Top 10.

GRAPE VARIETIES

There are many species of grapevines grown around the world, but wine is made with a single species called *Vitis vinifera*, native to Eurasia. The grape names we see on wine labels—like Chardonnay and Shiraz—are the names of grape varieties within this vinifera species. Grape varieties are like breeds of dogs: All are members of one species, but they have been bred for specific characteristics. In grapes, varieties may be bred for eating fresh or winemaking, for red wine or white wine, to survive a region's harsh winters or another's summer droughts.

WORKHORSE GRAPES

There are thousands of grape varieties, but only a few dozen are of commercial significance in winemaking. Of these, a short list of 10 represents the vast majority that wine drinkers are likely to encounter on labels. These are not the 10 most widely planted grapes on earth; those are often workhorse grapes, like Airén and Cinsaut, used for making bulk wine or brandy. And the list excludes many grapes that make truly exceptional wine and are popular in their home region but haven't caught on elsewhere, such as Tempranillo and Nebbiolo.

TOP 10 GRAPE VARIETIES

This Top 10, in no particular order, includes just the most famous grapes—the wine world's jet-set celebrities. They are the grapes we see most often on labels because they are wildly popular and planted all over the world, not just on their home turf but abroad, too.

WHITE WINE GRAPES

- Chardonnay
- Sauvignon Blanc
- Riesling
- Pinot Grigio/Pinot Gris
- Moscato

RED WINE GRAPES

- Cabernet Sauvignon
- Merlot
- Pinot Noir
- Syrah/Shiraz
- Grenache/Garnacha

MORE FAMOUS OVERSEAS

Why do European grapes earn more name recognition when they're grown in the New World than they do at home? The traditional convention for naming wines in Europe is to use the place name. The type of grape used is not always mentioned. In the Americas and southern hemisphere, printing the grape name on the label is the norm. Sangiovese may be the number one grape in Italy, but its name isn't as familiar as the names of its most famous wines, like Chianti or Brunello di Montalcino. Meanwhile, Malbec is a household name simply because vintners in Argentina, where most is grown, place it on the label.

WHY GRAPES?

Wine can be made from all sorts of fruit, but grapes are the fruit of choice because they have the sweetest juice. Wine grapes contain at least 20 percent fermentable sugar when ripe and over 70 percent water, which makes them ideal for making into an alcoholic drink.

JUICY FRUIT
Every part of the grape is used in making red and rosé wines. For white wines, only the juice is used, and all solids are discarded.

STEM
The grape stalk is usually removed and discarded but may be used in the fermentation of some red wine styles.

SKIN
The peel is where color compounds and tannins are found, making it essential in red-wine making. This is also where most flavor compounds are located, or in the flesh immediately underneath.

SEEDS
Vintners are careful not to crush the pips, because they can release bitter flavors into wine.

JUICY FLESH
The transparent pulp contains three of the main components for winemaking: fermentable sugar, refreshing acidity, and plenty of water.

CHARDONNAY

No other grape can match the popularity of Chardonnay, which can make some of the richest, strongest white wines on earth without losing finesse. Chardonnays are almost always dry in style, and only mass-market brands flirt with a hint of sweetness. The vine adapts well to varying climates, so every country that produces wine grows at least a little Chardonnay.

GREEN GRAPE FOR WHITE WINES
APPLEY FLAVOR FAMILY

THE FLAVOR RANGE

CRAB APPLE · GREEN APPLE · PINEAPPLE · APPLE PIE

LOW RIPENESS/ COOL CLIMATE

HIGH RIPENESS/ WARM CLIMATE

SEASONED WITH OAK

Chardonnay is very subtle in flavor, so vintners often amplify its aromatics by fermenting in new barrels. Grapes must be fully ripe for oak to be a pleasing accent, however, so its presence tends to be strongest in the heaviest wines from the warmest regions. New World Chardonnays are so often oaky, for example, that many wine drinkers think their toasty cognac-like flavors are inherent to the grape, but this is not so. Unoaked Chardonnays, sometimes called "unwooded," can taste as crisp and mild as Pinot Grigio.

THE CHARDONNAY STYLE RANGE

Wines from this grape fall in the sensory trait ranges shown below.

ATTRIBUTE	RANGE
COLOR	White only
COLOR DEPTH	Full range: from pale to dark
SWEETNESS	Partial range: from dry to lightly sweet
ACIDITY	Full range: from low acid to tart
FRUIT INTENSITY	Partial range: from mild to flavorful
OAK PRESENCE	Full range: from none to strong oak
WEIGHT/BODY	Full range: from light to heavy
CARBONATION	Full range: from still to sparkling

RED-LIKE WHITE

Red wines are typically heavier and oakier than whites, but Chardonnays can compete on both counts. Not only are they often oaked (unlike most white wines), but they feel richer for another reason, too. Chardonnay can produce more sugar than other green grapes without losing refreshing acidity, and this boosts the alcohol content of its wines.

WHERE'S IT FROM?

Chardonnay originated in France's Burgundy region, where quality viticulture developed in medieval times. Even today, all white wines from Burgundy are 100 percent Chardonnay. Classy, understated "white Burgundy" remains the benchmark Chardonnay style, emulated by vintners worldwide.

Warmer New World regions like California and Australia grow riper Chardonnay, making bold, fruity, opulent wines. These are tremendously popular and conveniently labeled with the grape name. Cooler regions, like New Zealand and Canada, make lighter, brighter, more Burgundian styles.

Paris

Burgundy

FRANCE

ARCHETYPES:

FRENCH WHITE BURGUNDY

Typically unoaked in coolest regions and modest appellations: Bourgogne Blanc, Mâcon-Villages, Chablis, St-Véran, Viré-Clessé

Typically barrel-fermented in warmer premium appellations: Meursault, Puligny-Montrachet, Pouilly-Fuissé, Chassagne-Montrachet

Chardonnay is also one of three grapes grown in neighboring Champagne: *Blanc de blancs* is 100 percent Chardonnay.

Burgundy

OTHER TOP CHARDONNAY REGIONS

1 US	California: Sonoma, Santa Barbara, Monterey Other states: Washington, Oregon, New York	
2 AUSTRALIA	South Australia: Adelaide, Padthaway Other states: Victoria, New South Wales, Western Australia	
3 CHILE	Casablanca Valley, Maipo Valley, Aconcagua	
4 SOUTH AFRICA	Coastal Region, Stellenbosch, Cape South Coast	
5 NEW ZEALAND	Hawke's Bay, Gisborne, Marlborough	
6 CANADA	Niagara Peninsula, Okanagan Valley	

THE GOLDEN CHILD OF WINE
Winemakers love Chardonnay for many reasons. The vines adapt well to diverse climate conditions, the golden-tinged grapes make excellent wine in a variety of different styles, and the Chardonnay name is trusted by wine drinkers worldwide.

SAUVIGNON BLANC

Wines made from Sauvignon Blanc are more recognizable than most whites, thanks to their sharp acidity and distinctive pungent scent, typically of green foods such as herbs, vegetables, or even green fruit—from lime to honeydew melon. These wines are tremendously popular, largely because they deliver reliably good value at reasonable prices.

**GREEN GRAPE
FOR WHITE WINES**
HERBAL FLAVOR FAMILY

THE FLAVOR RANGE

CRAB APPLE · GRAPEFRUIT · PASSION FRUIT · KIWI FRUIT

**LOW RIPENESS/
COOL CLIMATE**

**HIGH RIPENESS/
WARM CLIMATE**

TWIN STYLES

Most Sauvignon Blanc wines are medium-bodied and dry, but they can be made in one of two distinct styles. The most common is the cool-climate Loire model, popularized in New Zealand: sharp and citrusy unoaked wines, with mouthwatering acidity. Where greater ripeness is possible, such as in California, the white Bordeaux model is sometimes followed for premium wines. Using oak and blending with Semillon enriches and softens this "savage" grape to give wines that are plumper, less "green" in flavor, and less aggressive in their acidity.

THE SAUVIGNON BLANC STYLE RANGE

Wines from this grape fall in the sensory trait ranges shown below.

ATTRIBUTE	RANGE
COLOR	White only
COLOR DEPTH	Pale only
SWEETNESS	Dry only
ACIDITY	Partial range: from tangy to tart
FRUIT INTENSITY	Partial range: from flavorful to bold
OAK PRESENCE	Partial range: from none to mild oak
WEIGHT/BODY	Partial range: from light to mid-weight
CARBONATION	Partial range: from still to spritzy

WILD VINE

The name Sauvignon may derive from the French word *sauvage* (meaning untamed or savage) and might refer to this grape's resemblance to wild vines or possibly to the ferocity of its scent. Either way, the name was passed on to its offspring, the famed red Cabernet Sauvignon grape.

WHERE'S IT FROM?

Sauvignon Blanc is indigenous to the upper reaches of the Loire Valley south of Paris. The region's tart, unoaked white wines from appellations like Sancerre and Pouilly-Fumé serve as the inspiration for most international variants, like New Zealand Sauvignon Blanc. However, this grape has also been grown for centuries in Bordeaux on France's Atlantic coast, where it is blended with other local grapes. White Bordeaux are often richer and the finest examples may be barrel-fermented, an approach to Sauvignon Blanc that has been embraced in the US.

ARCHETYPES:

FRENCH LOIRE WHITES AND WHITE BORDEAUX

Loire Valley: 100 percent Sauvignon Blanc
Typically unoaked and bottled young in all appellations: Sancerre, Pouilly-Fumé, Touraine

Bordeaux: Sauvignon Blanc-based blends with Semillon
Typically unoaked and bottled young in modest appellations: Bordeaux, Entre-Deux-Mers

Typically barrel-fermented and aged in premium appellations: Graves, Pessac-Léognan

Loire Valley and Bordeaux

OTHER TOP SAUVIGNON BLANC REGIONS

1	NEW ZEALAND	South Island: Marlborough, Canterbury North Island: Hawke's Bay, Gisborne
2	US	California: Sonoma, Napa, Central Coast Washington: Columbia Valley
3	SOUTH AFRICA	Coastal Region, Cape South Coast
4	CHILE	Casablanca Valley, Maipo Valley
5	ITALY	Trentino, Alto Adige, Friuli

PROLIFIC AND FLAVORFUL

Sauvignon Blanc is an uncommonly vigorous vine that can yield bumper crops without losing much flavor intensity. However, its aromatics can tilt toward the vegetable realm when too much of its energy is spent producing shoots and leaves.

RIESLING

If Riesling has a superpower, it is the ability to make flavorful, balanced wines at freakishly low alcoholic strength. Vintners often make pleasing sweet wines from Riesling as a result, since it is less essential to convert grape sugar into alcohol than for other varieties. These wines have become Riesling's claim to fame because they succeed at something other grapes fail to do well.

GREEN GRAPE FOR WHITE WINES
APPLEY FLAVOR FAMILY

THE FLAVOR RANGE

LIME   GREEN APPLE   PEACH   APRICOT

**LOW RIPENESS/
COOL CLIMATE**

**HIGH RIPENESS/
WARM CLIMATE**

HOW SWEET IS IT?

The most popular and most widely known style of Riesling takes its inspiration from the lightly sweet wines of Germany's Mosel and Rheingau regions—wines that feature the sweet-tart balance of green apples. However, Riesling can be found at every degree of sweetness—from bone-dry, to candy sweet dessert wines—both inside Germany and in other parts of the world. Dry interpretations model themselves on the French Alsace style, where Riesling makes graceful wines that are fuller-bodied and much less sweet.

THE RIESLING STYLE RANGE

Wines from this grape fall in the sensory trait ranges shown below.

ATTRIBUTE	RANGE
COLOR	White only
COLOR DEPTH	Partial range: from pale to moderate
SWEETNESS	Full range: from dry to fully sweet
ACIDITY	Partial range: from tangy to tart
FRUIT INTENSITY	Full range: from mild to bold
OAK PRESENCE	None
WEIGHT/BODY	Full range: from light to heavy
CARBONATION	Partial range: from still to spritzy

BUILT FOR AGING

Wines made from Riesling have exceptional aging potential and can improve in the cellar for decades longer than most white wines. People often assume that lighter whites need to be drunk young, but Riesling has a strong natural resistance to oxidation, thanks in part to its high levels of acidity.

WHERE'S IT FROM?

Riesling is native to the southwestern German valleys of the Rhine and its tributaries between Frankfurt and Trier, where it makes delicate wines that often feature a touch of natural grape sugar. Across the border in France, Riesling also has a long history in Alsace, where it traditionally makes a stronger, drier wine. This vine makes its best international wines in colder climates, too. Vintners making Riesling in North America and New Zealand most often follow the sweeter German style, while those in Austria and Australia tend to favor the drier French approach.

ARCHETYPES:

GERMAN RIESLING AND FRENCH ALSACE RIESLING

Germany: 100 percent Riesling
Traditionally light, tart, and lightly sweet—often below 11% alcohol—in top appellations: Mosel, Rheingau, Pfalz, Rheinhessen

France: 100 percent Riesling
Traditionally mid-weight, tart, and dry—usually over 12.5% alcohol—in Alsace

Rhine Valley and Alsace

OTHER TOP RIESLING REGIONS

1 US	Washington: Columbia Valley New York: Finger Lakes Other states: California, Oregon	
2 AUSTRIA	Lower Austria, Vienna, Burgenland	
3 AUSTRALIA	South Australia: Clare Valley, Eden Valley Other states: Victoria, Western Australia, Tasmania	
4 CANADA	Niagara Peninsula, Okanagan Valley	
5 NEW ZEALAND	South Island: Marlborough, Otago, Nelson	

SMALL AND INTENSE

Riesling grapes are very small, which increases the flavor intensity of its wines, since the flesh closest to the berry's skin is highest in aromatic compounds. Compared to other fine-wine grapes, it ripens very early and produces healthy yields.

PINOT GRIGIO/PINOT GRIS

This is an odd grape in many ways. Not only is it a red grape that makes white wine under two different names, but its wines have a split personality. Under its French identity, Pinot Gris makes wines that are opulent and flavor-rich; while as Pinot Grigio from northern Italy, it makes one of the most popular wines in the world in a much lighter, much milder style.

**PURPLE GRAPE
FOR WHITE WINES**
APPLEY FLAVOR FAMILY

THE FLAVOR RANGE

GREEN PEAR RED APPLE PEACH CANTALOUPE

**LOW RIPENESS/
COOL CLIMATE**

**HIGH RIPENESS/
WARM CLIMATE**

THE GRAY PINOT

As its name suggests, this grape is related to Burgundy's red Pinot Noir. It is a paler-skinned mutation that looks pinkish red rather than deep purple—hence its original name Pinot Gris: gray Pinot. Nowadays, though, we more often see it labeled in Italian: Pinot Grigio. In the New World, the name used often reflects the wine's style.

Vintners using the lighter Italian model and name harvest early to retain freshness. Those labeling their wine Pinot Gris are more likely to let the grapes ripen more fully, French-style, making stronger and more fragrant wine.

THE PINOT GRIGIO/PINOT GRIS STYLE RANGE

Wines from this grape fall in the sensory trait ranges shown below.

ATTRIBUTE	RANGE
COLOR	Partial range: from white to pink
COLOR DEPTH	Full range: from pale to dark
SWEETNESS	Partial range: from dry to lightly sweet
ACIDITY	Full range: from low acid to tart
FRUIT INTENSITY	Partial range: from mild to flavorful
OAK PRESENCE	Partial range: from none to mild oak
WEIGHT/BODY	Full range: from light to heavy
CARBONATION	Partial range: from still to spritzy

BRASSY ROSÉ

Pinot Grigio grapes are technically red, but they lack the color saturation needed to make satisfying red wines so are typically made into white wine instead. However, if the grapes are fermented with full skin contact, they can make a coppery pink rosé.

WHERE'S IT FROM?

Originally from Burgundy in France, Pinot Gris migrated east and settled comfortably in Alsace, where it still makes rich, plump wines. From there, it traveled across Germany and into the Dolomites in what is now northern Italy. It was common practice to harvest grapes early in these valleys, at lower ripeness than in sun-drenched Alsace, and the resulting wines were much lower in alcohol and milder in flavor. When the vine spread to the plains around Venice, this early-harvest approach remained the norm for Italian Pinot Grigio because it was easy-drinking and inexpensive to produce.

ARCHETYPES:

ITALIAN PINOT GRIGIO AND FRENCH ALSACE PINOT GRIS

Italy: 100 percent Pinot Grigio
Light, mild, and crisp "low ripeness" style—usually below 13% alcohol—in top appellations: Venezie, Trentino, Friuli

France: 100 percent Pinot Gris
Rich, aromatic, and plump "high ripeness" style—usually over 13% alcohol—in Alsace

Northern
Italy and Alsace

OTHER TOP PINOT GRIGIO/PINOT GRIS REGIONS

1 US	California: Sonoma, Central Coast Oregon: Willamette Valley	
2 GERMANY	Pfalz, Baden	
3 CANADA	Niagara Peninsula, Okanagan Valley	
4 AUSTRIA	Styria, Burgenland	
5 NEW ZEALAND	Hawke's Bay, Gisborne, Marlborough	
6 AUSTRALIA	South Australia, Victoria	

A FADED GRAPE
Two genes in grapes produce the color compounds that result in purple skins. Pinot Gris is a mutation of Pinot Noir in which one of these has been deactivated, resulting in its odd faded purple coloration.

MOSCATO

A diverse range of wine styles are made from Moscato, but whether they're fresh and fruity or dark and raisiny, almost all are sweet and smell of flowers. The perfumed scent derives from terpenes—aroma compounds that cause Moscato to be markedly more pungent than other wine styles (though similarly terpene-rich Gewurztraminer comes close).

**GREEN GRAPE
FOR WHITE WINES**
FLORAL FLAVOR FAMILY

THE FLAVOR RANGE

GREEN GRAPE · LYCHEE · ROSEWATER · DRIED APRICOT

**LOW RIPENESS/
COOL CLIMATE**

**HIGH RIPENESS/
WARM CLIMATE**

FROM JUICY TO RAISINY

The aromatic intensity of Moscato can seem unbalanced in dry wines, so it tends to be used to make fully sweet styles. The most famous Moscatos are made by using either temperature control or additional distilled spirit to interrupt fermentation, as with the low-alcohol, bubbly Moscatos of Asti and the high-alcohol, liqueur-like Muscats of the south of France. However, in many sunny regions—such as Spain, Portugal, and Australia—the grapes are dried before winemaking begins, to make a dark, thick, and sticky fortified dessert wine.

THE MOSCATO STYLE RANGE

Wines from this grape fall in the sensory trait ranges shown below.

ATTRIBUTE	RANGE
COLOR	Full range: from white to red
COLOR DEPTH	Full range: from pale to dark
SWEETNESS	Full range: from dry to fully sweet
ACIDITY	Full range: from low acid to tart
FRUIT INTENSITY	Bold only
OAK PRESENCE	None
WEIGHT/BODY	Full range: from light to heavy
CARBONATION	Full range: from still to sparkling

MUSKY NAMES

Moscato is known as Muscat in French and Moscatel in Spanish, and its name can even be an adjective when variants of other grapes display a floral scent, as with Chardonnay Musqué. But the similar-sounding Muscadet, a white from northern France, is confusingly of no relation and bears no aromatic resemblance whatsoever.

WHERE'S IT FROM?

Moscato's fine-wine potential was already well known in ancient Greece, and it may be a common ancestor of many top grapes. It is not a single variety but, rather, a family of grapes with more than 100 distinct variants, many cultivated for table grapes or raisins throughout the Mediterranean. While most Moscato varieties are white, some are purple and can be used to make red wine. Wines made from these fragrant grapes are common in much of southern Europe but are also found in the New World.

OTHER TOP MOSCATO REGIONS

1 FRANCE	Rhône, Languedoc-Roussillon (sweet), Alsace (dry)
2 PORTUGAL	Setúbal, Douro
3 AUSTRALIA	Victoria: Rutherglen Other states: New South Wales, South Australia
4 US	California
5 GREECE	Samos, Patras, Rhodes

VINEYARDS ALL ABUZZ
The name Moscato may seem to reference musk, the perfume ingredient. However, ancient writings suggest the name refers instead to the insects that are attracted to this vine's fragrant fruit as it ripens—flies belong to the genus known as *Musca*.

ARCHETYPES:

ITALIAN MOSCATO

Piedmont, Italy: 100 percent Moscato Bianco
Low alcohol, sweet, sparkling, and fresh tasting—under 10% alcohol—in Asti

ARCHETYPES:

SPANISH MOSCATEL

Andalusia, Spain: 100 percent Muscat of Alexandria
High alcohol, sweet, fortified, and from dried grapes—over 15% alcohol—in Sherry and Malaga

CABERNET SAUVIGNON

Cabernet Sauvignon is the world's number one wine grape by planted area. Compared to its rivals, it has unusually small berries and thick skins, resulting in more purple solids and less clear juice. In red wines, more dark grape skins result in more color, flavor, and tannins, so these traits are very intense in Cabernet Sauvignon wines, especially in those from low-yielding premium vineyards.

PURPLE GRAPE FOR RED WINES

BLACK-FRUIT FLAVOR FAMILY

THE FLAVOR RANGE

ROASTED PEPPER · BLACK CURRANT · BLACKBERRY · BRANDIED CHERRY

LOW RIPENESS/ COOL CLIMATE

HIGH RIPENESS/ WARM CLIMATE

PERFECT BLENDING GRAPE

Rich in natural preservatives, Cabernet Sauvignon can seem harsh when young, but its resistance to oxidation is a blessing for age-worthy premium wines. Pure Cabernet Sauvignon's intensity in color, flavor, and tannin can yield lean, aggressive wines, but it is perfectly suited to blending with milder grapes.

When it takes a dominant role, Cabernet Sauvignon is often softened by adding fruitier wine from fleshier grapes like Merlot or Shiraz. In smaller proportions, it can add density and aging potential to wines made with thinner-skinned grapes, such as Merlot, Sangiovese, or Malbec.

THE CABERNET SAUVIGNON STYLE RANGE

Wines from this grape fall in the sensory trait ranges shown below.

ATTRIBUTE	RANGE
COLOR	Red only
COLOR DEPTH	Partial range: from moderate to dark
SWEETNESS	Dry only
ACIDITY	Partial range: from tangy to tart
FRUIT INTENSITY	Bold only
OAK PRESENCE	Full range: from none to strong oak
WEIGHT/BODY	Partial range: from mid-weight to heavy
TANNIN	Partial range: from velvety to leathery

MYSTERY BLENDS

Some of Cabernet Sauvignon's top wines don't list its name on the front label. Not only do European classics like Bordeaux and Bolgheri rarely mention their grapes, but international blends are not permitted to name a grape unless the wine contains at least 75 percent of that variety.

WHERE'S IT FROM?

Cabernet Sauvignon is indigenous to Bordeaux, where it has historically dominated the blend in the most prestigious wines of France's premier red wine region. Cabernet Sauvignon's flavor density and capacity to age gracefully have earned it a role in the top red wines of many nations. In the New World, it produces the most expensive and collectible red wines made in the US, Chile, and South Africa, which are often Bordeaux-inspired blends.

FRANCE

Paris

Bordeaux

ARCHETYPE:

FRENCH RED BORDEAUX

Dark, intense blends, with strong oak and high tannin

Typically dominates blends in the warmest premium appellations: Haut-Médoc, Margaux, Pauillac, St-Estèphe, St-Julien, Moulis

Typically has a supporting role in cooler modest appellations: Bordeaux, Graves, St-Émilion

Bordeaux

1   2

3   5   4

OTHER TOP CABERNET SAUVIGNON REGIONS

1 US	California: Napa Valley, Sonoma, Paso Robles Washington: Columbia Valley	
2 ITALY	Tuscany, Trentino	
3 CHILE	Maipo Valley, Rapel Valley, Aconcagua	
4 AUSTRALIA	South Australia: Coonawarra	
5 SOUTH AFRICA	Coastal Region, Stellenbosch, Paarl	

SUN-LOVING VARIETY
Cabernet Sauvignon vines bud later in the spring than most Bordeaux varieties, and their fruit takes longer to ripen in the fall. Their small, thick-skinned berries need ample sun to develop to their full potential in color and flavor.

MERLOT

If Bordeaux's Cabernet Sauvignon is a muscular superhero, its close relation Merlot is its lithe sidekick with a sexier name. Wines made from this variety are not as dark, tannic, or intense as those of Cabernet Sauvignon as a rule. They are softer and milder, with more voluptuous fruit flavors—characteristics that make Merlot more appealing for drinking young.

PURPLE GRAPE FOR RED WINES

BLACK-FRUIT FLAVOR FAMILY

THE FLAVOR RANGE

TOMATO · BLACK PLUM · BLACKBERRY · CHERRY PIE

LOW RIPENESS/ COOL CLIMATE

HIGH RIPENESS/ WARM CLIMATE

THE PRICE OF POPULARITY

Merlot is so crowd-pleasing that it is often exploited for volume in everyday wines, and its reputation has been devalued a little by cheap bottlings. However, when planted in top vineyard sites and made to rigorous standards, Merlot makes wines of incredible power and grace. In Bordeaux, this grape is typically relegated to marginal land; however, cult wines made with 100 percent Merlot, like the legendary Château Pétrus, command higher prices than their Cabernet-based rivals, proving that this variety is one of the world's finest red grapes.

THE MERLOT STYLE RANGE

Wines from this grape fall in the sensory trait ranges shown below.

ATTRIBUTE	RANGE
COLOR	Red only
COLOR DEPTH	Partial range: from moderate to dark
SWEETNESS	Dry only
ACIDITY	Full range: from low acid to tart
FRUIT INTENSITY	Partial range: from flavorful to bold
OAK PRESENCE	Full range: from none to strong oak
WEIGHT/BODY	Partial range: from mid-weight to heavy
TANNIN	Full range: from none to leathery

DRINK ME NOW

Merlot often gets overshadowed by Cabernet Sauvignon because its wines are lighter, softer, and do not age as gracefully. But in a world where instant gratification is the name of the game, Merlot's supposed weaknesses are really its strengths.

WHERE'S IT FROM?

Merlot and Cabernet Sauvignon are closely related grapes from Bordeaux that share a family resemblance in aromatics but differ in their power, color, and tannic grip. In Bordeaux, Merlot is by far the most widely planted of the two because it ripens much earlier and is therefore a more economically viable variety. However, this grape truly comes into its own in the Americas, making luscious, friendly red wines labeled under its own name, from Chile in the south, to California and Washington in the north.

FRANCE

Paris

Bordeaux

ARCHETYPE:

FRENCH RED BORDEAUX

Mid-weight, flavorful blends with soft tannin and mild oak

Typically dominates blends in cooler modest appellations: Bordeaux, Graves, St-Émilion, Pomerol

Typically plays a supporting role in the warmest premium appellations: Haut-Médoc, Margaux, Pauillac, St-Estèphe, St-Julien, Moulis

Bordeaux

OTHER TOP MERLOT REGIONS

1	US	California: Napa Valley, Sonoma Washington: Columbia Valley
2	CHILE	Maipo Valley, Rapel Valley, Aconcagua
3	ITALY	Northern Italy, Tuscany
4	NEW ZEALAND	Hawke's Bay, Marlborough
5	CANADA	Okanagan Valley, Niagara Peninsula

FRAGILE BUT GENEROUS
Merlot's thinner skins and looser clusters make it more fragile in the vineyard than other Bordeaux varieties like Cabernet Sauvignon, but this disadvantage is more than offset by its early ripening and generous yields per vine.

PINOT NOIR

In many ways, Pinot Noir is the polar opposite of Cabernet Sauvignon. Pinot Noir has thin skins, not thick, and makes much lighter, paler wines that are not as well suited to blending or long-term aging. Cabernet Sauvignon needs extra heat to ripen fully, while Pinot Noir needs cooler conditions to retain its seductive earthy charm and falls flat in places where it gets too ripe and fruity.

**PURPLE GRAPE
FOR RED WINES**
RED-BERRY FLAVOR FAMILY

THE FLAVOR RANGE

STRAWBERRY · POMEGRANATE · RASPBERRY · CHERRY

**LOW RIPENESS/
COOL CLIMATE**

**HIGH RIPENESS/
WARM CLIMATE**

SEDUCTIVE FINESSE

Pinot Noir is Cabernet Sauvignon's closest rival for the red wine crown—indeed, many feel it makes superior wines. While Cabernet Sauvignon can perform reliably in many wine regions and sets itself apart with sheer power, Pinot Noir is much more fickle and seduces with its finesse. This seemingly weaker red wine only truly shines in a handful of regions, and it can disappoint when conditions aren't just right. But when Pinot Noir excels, it makes wines that are hauntingly beautiful and speak directly to the soul in a way that Cabernet Sauvignon cannot.

THE PINOT NOIR STYLE RANGE

Wines from this grape fall in the sensory trait ranges shown below.

ATTRIBUTE	RANGE
COLOR	Red only
COLOR DEPTH	Partial range: from pale to moderate
SWEETNESS	Dry only
ACIDITY	Full range: from low acid to tart
FRUIT INTENSITY	Partial range: from flavorful to bold
OAK PRESENCE	Full range: from none to strong oak
WEIGHT/BODY	Partial range: from mid-weight to heavy
TANNIN	Partial range: from none to velvety

FINGERPRINTING

The term *terroir* was coined in Burgundy to capture Pinot Noir's uncanny knack of tasting slightly different in every place it's grown. Centuries ago, it was noted that each aspect of a vineyard—from soil type to orientation of its slope—could create a recognizable flavor, a unique fingerprint of the land, captured in its wine.

WHERE'S IT FROM?

Pinot Noir is, in one sense, the original wine grape; it is the red grape of Burgundy, where fine wine as we know it was born in medieval France. Pinot Noir prefers cooler growing conditions than most red grapes and is too fragile for large-scale farming. Outside of Burgundy, it makes exceptional wine in only a few small regions like Oregon and California in the US, New Zealand and southeastern Australia, and Germany and northern Italy.

Paris ○

Burgundy ----------○

FRANCE

ARCHETYPE:

FRENCH RED BURGUNDY

All Burgundy red wines except Beaujolais are 100 percent Pinot Noir

Typically light, pale, and tart in modest appellations: Bourgogne, Mercurey, and Côte de Beaune villages like Santenay and Chorey-lès-Beaune

Richer, darker, and more fragrant in premium appellations: most Côte de Nuits villages like Gevrey-Chambertin, Nuits-St-Georges, and select Côte de Beaune villages like Volnay and Pommard

Pinot Noir is also one of three grapes grown in neighboring Champagne.

Burgundy

OTHER TOP PINOT NOIR REGIONS

1 US	California: Sonoma, Santa Barbara, Monterey Oregon: Willamette Valley
2 GERMANY	Pfalz, Baden, Rheingau
3 NEW ZEALAND	Central Otago, Martinborough, Marlborough
4 ITALY	Trentino-Alto Adige
5 AUSTRALIA	Yarra Valley, Adelaide Hills, Tasmania
6 SOUTH AFRICA	Franschhoek, Walker Bay, Cape South Coast
7 CANADA	Niagara Peninsula, Okanagan Valley

HARD WORK PAYS OFF
Pinot Noir is an early-ripening variety that performs best in cooler regions where its final rush to ripeness is slowed. The finest variants of this cultivar are low-yielding and demanding to farm, but the results are well worth the effort.

SYRAH/SHIRAZ

Whether it's called by its original French name Syrah or its southern hemisphere alias Shiraz, this Rhône grape is a true powerhouse. Its wines can rival Cabernet Sauvignon for intensity, depth, and aging capacity because the variety has similarly small berries with thick, dark skins. Syrah needs ample sun to reach full ripeness but has enough juicy fruit character to stand alone easily.

**PURPLE GRAPE
FOR RED WINES**
SPICED-FRUIT FLAVOR FAMILY

THE FLAVOR RANGE

RASPBERRY   BLACK PLUM   BLACKBERRY JAM   BLUEBERRY PIE

**LOW RIPENESS/
COOL CLIMATE**

**HIGH RIPENESS/
WARM CLIMATE**

FLAVOR AND COLOR BOOSTER

Syrah's distinctive spicy scent, blue-tinged color, and preservative tannins make it a marvelous blending partner for intensifying milder or paler grapes. While it can taste delicious and balanced on its own, many vintners soften its impact with small amounts of other grapes. This pattern is inspired by Syrah's traditional role in the Rhône, where most is used to bolster paler Grenache, as with Côtes du Rhône. In smaller appellations, it is often mellowed with a smidgen of local green grapes. Syrah is more generally seen unblended in the New World, often as Shiraz.

THE SYRAH/SHIRAZ STYLE RANGE

Wines from this grape fall in the sensory trait ranges shown below.

ATTRIBUTE	RANGE
COLOR	Red only
COLOR DEPTH	Partial range: from moderate to dark
SWEETNESS	Partial range: from dry to lightly sweet
ACIDITY	Full range: from low acid to tart
FRUIT INTENSITY	Partial range: from flavorful to bold
OAK PRESENCE	Full range: from none to strong oak
WEIGHT/BODY	Partial range: from mid-weight to heavy
TANNIN	Full range: from none to leathery

PEPPER POT

Many Syrah wines have an uncanny smell reminiscent of pepper. In chilly regions and at low ripeness, it can come across as a pickled green-peppercorn scent, but with more sunshine it develops into a cracked black pepper smell that can make you want to sneeze. However, at the very highest ripeness levels, this aroma becomes less prominent.

WHERE'S IT FROM?

Syrah hails from the south of France, where its primary use is as a *vin médecin*, strengthening blends in the southern Rhône Valley. Syrah does make its own wine in a few tiny appellations farther north, along the Rhône's steep banks, like tart Crozes-Hermitage and spicy Côte-Rôtie, but these are rare, and many are quite expensive. As a result, it is in jammier New World wines that Syrah takes center stage, most notably as Australia's signature grape Shiraz.

FRANCE

Paris

Northern Rhône Valley

ARCHETYPE:

FRENCH NORTHERN RHÔNE REDS

Northern Rhône Valley Syrah-based wines—80–100 percent; dark, tannic red wines with intense peppery aromatics

Modest appellations: Crozes-Hermitage, St-Joseph

Premium appellations: Hermitage, Côte-Rôtie, Cornas

Rhône Valley

OTHER TOP SYRAH/SHIRAZ REGIONS

1	AUSTRALIA	South Australia: Barossa, McLaren Vale Other states: Victoria, New South Wales, Western Australia
2	US	California: Napa Valley, Sonoma, Santa Barbara Washington: Columbia Valley
3	SOUTH AFRICA	Coastal Region, Paarl, Stellenbosch
4	CHILE	Maipo Valley, Rapel Valley, Aconcagua
5	SPAIN	Castilla-La Mancha, Catalonia
6	CANADA	Okanagan Valley

PERSIAN MYTH-BUSTING

Syrah's alternative moniker, Shiraz, has led many to conclude that the vine originally came from the Iranian city of Shiraz, where grapes are grown more for raisins than for wine. However, recent DNA testing has revealed that this grape hails from southeastern France.

GRENACHE/GARNACHA

The Rhône Valley's workhorse grape can make affordable easy-drinking red wines, as well as sappy, quaffable rosés. But it can also produce rich, concentrated premium reds with savory spiced aromatics and quite high alcohol levels. Since Grenache oxidizes easily, quickly turning rusty orange, vintners often dose it with dashes of darker Rhône grapes like Syrah or Mourvèdre as a natural preservative.

PURPLE GRAPE FOR RED WINES

SPICED-FRUIT FLAVOR FAMILY

THE FLAVOR RANGE

WATERMELON RED CHERRY STRAWBERRY JAM ROASTED FIGS

LOW RIPENESS/ COOL CLIMATE

HIGH RIPENESS/ WARM CLIMATE

WIDELY PLANTED BUT RARELY NAMED

Grenache wines feature ripe red-fruit flavors, like strawberry jam, often with accents of white pepper and cured ham. Outside its native Spain, where it is called Garnacha, this grape is rarely named on labels. Many of Grenache's most famous French wines don't mention grapes by tradition, and its least ambitious wines go by generic brand names.

Most Grenache wines are blends, often with Rhône partners Syrah and Mourvèdre, so they often fall short of the 75 percent requirement for labeling by grape.

THE GRENACHE/GARNACHA STYLE RANGE

Wines from this grape fall in the sensory trait ranges shown below.

ATTRIBUTE	RANGE
COLOR	Partial range: from pink to red
COLOR DEPTH	Partial range: from pale to moderate
SWEETNESS	Partial range: from dry to lightly sweet
ACIDITY	Partial range: from low acid to tangy
FRUIT INTENSITY	Partial range: from flavorful to bold
OAK PRESENCE	Full range: from none to strong oak
WEIGHT/BODY	Full range: from light to heavy
TANNIN	Partial range: from none to velvety

THINK PINK

In addition to its rock-star red status, Grenache is the world's premier grape for making dry pink wines, thanks to its snappy, fresh strawberry flavors. It is no coincidence that dry rosé wines are specialties of the grape's home turf in southern France, in zones like the Rhône and Provence, as well as northeastern Spain.

WHERE'S IT FROM?

This grape traces its roots to Spain's arid Aragón, where it is called Garnacha, but its most influential role is as Grenache in France's southern Rhône Valley. The majority of French Grenache wines are blends, most often with Syrah and Mourvèdre in Rhône wines like Châteauneuf-du-Pape and Côtes du Rhône, but also with other Spanish grapes like Carignane in Languedoc and Roussillon. Outside Europe, Grenache grows in California and South Africa but is most important in Australia, often in Rhône-style blends labeled GSM, for Grenache, Shiraz, and Mourvèdre.

Paris ○

FRANCE

Southern
Rhône
Valley

ARCHETYPE:
FRENCH RHÔNE BLENDS

Southern Rhône Valley Grenache-based blends—70 percent plus; rich-textured red wines with meaty, peppery flavors

Modest regional-level appellation: Côtes du Rhône

Premium village-level appellations: Châteauneuf-du-Pape, Gigondas, Vacqueyras

Dry mid-weight rosé wines with fresh berry flavors

Modest regional-level appellation: Côtes du Rhône

Premium village-level appellation: Tavel (rosé only)

2
1
Rhône Valley

4

5   3

OTHER TOP GRENACHE/GARNACHA REGIONS

1 SPAIN	Catalonia, Aragón, Navarra
2 OTHER FRANCE	Languedoc, Roussillon, Provence
3 AUSTRALIA	South Australia: McLaren Vale, Barossa Other states: New South Wales, Victoria
4 US	California
5 SOUTH AFRICA	Coastal Region

A MANY-CLUSTERED THING
Grenache is known for its natural tendency to produce many clusters that can be very large in size. Reversing the pattern of most red grapes, its wines often look paler when grown in warmer regions and darker in cooler zones.

MINOR FRENCH GRAPES

Due to fine wine's complex history (see p.168), grapes of French heritage wield disproportionate influence in the wine world compared to those from elsewhere. Indeed, 7 of the top 10 grapes highlighted so far are native to France and all of them make high-profile fine wines there. A handful of lesser French grapes have enough global prominence to be worth getting to know better, too, such as those listed here.

BOTH AT HOME AND ABROAD

France's grapes do not simply dominate its own vineyards. Almost all fine wines made outside Europe use grapes of French origin.

CHENIN BLANC (AKA STEEN)

Chenin Blanc is an ancient grape native to the Loire Valley that is also widely planted in South Africa. This underrated vine is cost-efficient and can tolerate both hot and cold climates. Chenin Blanc not only rivals Riesling's epic talent for making sweet, low-alcohol wines, but it also trumps even Chardonnay's remarkable stylistic range, making exceptional sparkling wines in the coldest regions and world-class dry, oaked whites in warmer ones.

GREEN GRAPE FOR WHITE WINES
APPLEY FLAVOR FAMILY

o Paris

o---- Loire Valley

FRANCE

ARCHETYPE

FRENCH VOUVRAY

Vouvray: 100% Chenin Blanc
Typically off-dry, lightweight, unoaked whites, but also *brut* sparkling wines and sweet, botrytised dessert wines.

Savennières: 100% Chenin Blanc
Bone-dry, mid-weight white wines.

VIOGNIER

Viognier (pronounced VEE-own-yay) was once one of France's most obscure grapes, but has risen to global prominence through its embrace in New World regions like Australia and California. Its wines are nearly as fragrant and floral as Moscato and share Chardonnay's potential for rich texture and affinity for barrel fermentation. Viognier also plays a quirky role as a minor blending ingredient in red wines based on Syrah (aka Shiraz).

GREEN GRAPE FOR WHITE WINES
FLORAL FLAVOR FAMILY

o Paris

FRANCE
Rhône Valley ----o

ARCHETYPE

FRENCH CONDRIEU

Condrieu: 100% Viognier
Heavyweight, dry white wines with heady floral scent, often oaked.

Côte Rôtie (red): <5% Viognier
Tiny addition of Viognier to these Syrah reds adds fragrance and helps deepen and stabilize their red color.

CABERNET FRANC

Cabernet Franc is an ancient grape native to the Loire Valley that still makes light, tart reds there, but is better known for its minor role in Bordeaux-style red blends and as one of the genetic "parents" of the world's top red wine grape, Cabernet Sauvignon. Varietal Cabernet Franc wines resemble those of its famous offspring, but tend to be lighter, paler, more acidic, and more herbal. In the New World, it can make richer reds that are very satisfying.

PURPLE GRAPE FOR RED WINES
BLACK-FRUIT FLAVOR FAMILY

o Paris

o---- Loire Valley

FRANCE

ARCHETYPE
FRENCH CHINON AND BORDEAUX

Chinon and Bourgueil:
100% Cabernet Franc
Pale, tart, lightweight reds.

Cabernet d'Anjou:
Mostly Cabernet Franc
Dry, lightweight rosé wines.

Bordeaux, St-Émilion, and Pomerol: Minor role in these dry red blends, typically less than 25%.

GAMAY

Gamay is so prolific and easy to grow that it almost supplanted Pinot Noir in 14th-century Burgundy. Luckily for wine lovers, Gamay was outlawed in most of the region in 1395 as "very bad and disloyal" since its lightweight, fruity wines were considered inferior. However, Gamay still thrives in the Beaujolais subregion, where granite soils deepen its color and flavor. Outside of France, Gamay is grown in Switzerland and Canada, as well as in the Australian state of Victoria.

PURPLE GRAPE FOR RED WINES
RED-BERRY FLAVOR FAMILY

o Paris

Burgundy -----o

FRANCE

ARCHETYPE
FRENCH BEAUJOLAIS

Beaujolais, Beaujolais-Villages: 100% Gamay
Lightweight, dry red wines with low tannin and fruity aromas.

St. Amour, Brouilly, Morgon, and Moulin-à-Vent: 100% Gamay
Ten so-called *cru villages* of Beaujolais make more serious, concentrated Gamay wines.

MALBEC (AKA CÔT)

Malbec was once widely planted in southwestern France but has been largely supplanted by other grapes since it does not ripen well here. It remains the main grape of Cahors and a fringe player in Bordeaux's red blends, but is better known globally as Argentina's top wine grape. While French Malbec wines tend to be thin and rustic, Argentine versions are richer, darker, and bolder in flavor, featuring aromas of roasted beets and violets.

PURPLE GRAPE FOR RED WINES
BLACK-FRUIT FLAVOR FAMILY

o Paris

FRANCE

o---- Bordeaux and Cahors

ARCHETYPE
FRENCH CAHORS

Cahors: Minimum 70% Malbec
Mid-weight, dry red wines with firm tannins and sharp acidity.

Bordeaux, Bourg, and Blaye: Minor role in these dry red blends, typically 5% or less.

THE BEST GRAPES OF ITALY AND SPAIN

Italy is the world's top wine producer by volume and Spain ranks number one in vineyard area. However, neither Italian nor Spanish grapes are as well known as French grapes, and only a tiny fraction of wine grapes grown in the New World trace their roots to Italy or Spain. Of those varieties that do have name recognition in the global wine marketplace, these five loom the largest.

FOCUS DILUTED BY DIVERSITY

Italy grows over 350 wine grapes commercially, but their sheer numbers make it hard for individual grapes to develop a following with a global audience.

ALBARIÑO (AKA ALVARINHO)

Albariño is Spain's finest white grape, native to the cool northwestern Galicia region. Its "white Rhine" name evokes Riesling, and there is certainly a resemblance of sorts; both grapes feature a delicate yet penetrating scent and a lingering finish. However, there is no genetic relation. Albariño is also grown in northern Portugal, where it is spelled Alvarinho, and is being increasingly planted in New World regions like California and Australia.

GREEN GRAPE FOR WHITE WINES
APPLEY FLAVOR FAMILY

Galicia

o Madrid

SPAIN

ARCHETYPE

SPANISH RÍAS BAIXAS ALBARIÑO

Rías Baixas Albariño:
100% Albariño
Light to mid-weight white wines that are dry and unoaked.

Note: Rías Baixas wines that do not specify the Albariño grape may feature other local varieties or blend multiple grapes.

PROSECCO (AKA GLERA)

Near Venice in northern Italy, the Prosecco hills gave their name to a local grape variety, and to the region's wildly popular fresh and fruity sparkling wines made from that variety. While most sparkling wines are made with grapes that are also used for still wines, Prosecco is overwhelmingly made into bubbly. In 2009, the Prosecco grape was formally renamed "Glera," which is an expedient means to protect the Prosecco name as an appellation, or legal region of origin.

GREEN GRAPE FOR WHITE WINES
APPLEY FLAVOR FAMILY

Northern Italy

o Rome

ITALY

ARCHETYPE

ITALIAN PROSECCO

Prosecco: Minimum 85% Glera Lightweight, mild-flavored sparkling wines, typically featuring a hint of sweetness.

NEBBIOLO

Nebbiolo is native to cool and foggy Piedmont and serves as Italy's answer to Pinot Noir in many ways. Both vines are difficult to grow and produce pale, earthy wines that are capable of aging gracefully. Nebbiolo's flavor complexity makes it hard to classify, combining red and black berries with a riot of non-fruit scents from rose petals to cardamom to asphalt. Unlike Pinot Noir, though, Nebbiolo has yet to produce world-class wines in other countries.

PURPLE GRAPE FOR RED WINES
SPICED-FRUIT FLAVOR FAMILY

ARCHETYPE

ITALIAN BAROLO AND BARBARESCO

Barolo: 100% Nebbiolo
Bone-dry, heavyweight red wines with earthy flavors and high levels of acidity, tannin, and alcohol

Barbaresco: 100% Nebbiolo
Similar to Barolo, but typically a little lighter and milder.

SANGIOVESE

Italy's number one red-wine grape is Sangiovese, though it represents only 8 percent of total vineyard area. It is the dominant grape of Tuscany, where most "rossos" are either pure Sangiovese or Sangiovese-based blends. These wines are quite tart, tannic, and earthy, with high-pitched flavors of sour red cherries, but their tendency to brown quickly leads vintners to blend in darker grapes. Sangiovese is rarely grown outside of Italy.

PURPLE GRAPE FOR RED WINES
RED-BERRY FLAVOR FAMILY

ARCHETYPE

ITALIAN CHIANTI AND BRUNELLO DI MONTALCINO

Chianti and Chianti Classico:
Minimum 70% Sangiovese
Bone-dry, mid-weight red wines with pale color and high acidity.

Brunello di Montalcino: 100% Sangiovese Grosso
Smaller appellation for fine wines made solely with a superior clone.

TEMPRANILLO (AKA TINTA RORIZ, ARAGONÊS)

Tempranillo is Spain's top fine-wine grape and the main ingredient in many of its best red wines. It has the high acids and tannins needed for long-term aging, and many top-tier examples are well matured before release. Tempranillo offers a charming, dark, black-fruit profile that is quite popular with red-wine drinkers, even among wines that are lighter in weight. Tempranillo is not grown in significant volume outside of Spain and Portugal.

PURPLE GRAPE FOR RED WINES
BLACK-FRUIT FLAVOR FAMILY

ARCHETYPE

SPANISH RIOJA AND RIBERA DEL DUERO

Ribera del Duero (minimum 75% Tempranillo) and Rioja (Tempranillo-based blends):
Mid-weight, oaky red wines with firm tannins and intense flavor.

Toro: 100% Tinta de Toro
Smaller appellation whose wines are made solely with a superior local clone of Tempranillo.

THE BEST OF THE REST

Beyond those highlighted previously, there are many more fascinating grape varieties for wine lovers to explore, such as those in the chart below. The countries with the greatest proliferation of native wine grapes are those where grapes have been grown the longest and where climate conditions are vine-friendly. For example, Italy is the single largest wine producer on earth and more than 300 grapes are cultivated there for wine. But since most wine zones grow their own local grapes, these varieties rarely earn global name recognition and are seldom planted very far from home.

	GRAPE	NATIVE REGION	FAMOUS APPELLATIONS
WHITE WINE GRAPES	Gewurztraminer	Pfalz, Germany	Alsace; California
	Grüner Veltliner	Lower Austria	Wachau; Kamtal
	Macabeo	Cataluña, Spain	Cava; Penedes; Rioja; Languedoc; Roussillon
	Malvasia	Monemvasia, Greece	Madeira; Italy; Spain; Portugal
	Pinot Blanc	Burgundy, France	Alsace; Germany; Italy
	Semillon	Bordeaux, France	Sauternes; Barsac; Australia
	Torrontés	La Rioja, Argentina	Mendoza; La Rioja
RED WINE GRAPES	Aglianico	Campania, Italy	Taurasi; Vulture
	Barbera	Piedmont, Italy	Alba; Asti; Langhe
	Carmenère	Bordeaux, France	Chile
	Corvina	Veneto, Italy	Valpolicella; Bardolino
	Dolcetto	Piedmont, Italy	Alba; Asti; Langhe; Dogliani
	Monastrell	Valencia, Spain	Bandol; Rhône Valley; Jumilla, Yecla; Alicante
	Montepulciano	Abruzzi, Italy	Abruzzo
	Negroamaro	Puglia, Italy	Salice Salentino; Copertino; Squinzano
	Nero d'Avola	Sicily, Italy	Sicilia
	Petite Sirah	Rhône-Alpes, France	California; Australia; Languedoc
	Petit Verdot	Bordeaux, France	Bordeaux; California; Spain
	Pinotage	South Africa	Stellenbosch; Paarl
	Zinfandel	Croatia	California; Puglia; southern Italy

YET MORE UNIQUE GRAPES

Spain, Portugal, and Greece, like Italy, have dozens of their own unique wine grapes, but these are rarely cultivated outside their native zones. For historical reasons, French grapes have the edge in New World vineyards.

TYPICAL WINE STYLE	ALSO KNOWN AS
Pungent, floral whites	Traminer
Light, tangy whites	
Dry sparkling or white wines	Viura, Macabeu
Nutty dessert wines & light whites	Malmsey
Light, tangy whites	Pinot Bianco
Can be dry or fully sweet	
Dry, aromatic white wines	
Strong, spicy reds	
Tart, mid-weight reds	
Strong, herbal reds	
Fruity, light to heavyweight reds	
Fruity, lightweight reds	
Dark, strong reds	Mourvèdre; Mataro
Tangy, mid-weight reds	
Dark, strong reds	
Tangy, mid-weight reds	Calabrese
Strong, tannic reds	Durif
Strong, tannic reds	Cabernet Gernischt
Strong, meaty reds	Hermitage
Strong, jammy reds	Primitivo

CHAPTER CHECKLIST

Here is a recap of some of the most important points you've learned in this chapter.

- The grape names on wine labels are all **varieties** within the *Vitis vinifera* species.
- Grapes are well suited to winemaking due to their **high sugar and water** content.
- **Chardonnay** adapts well to varying climates, so every wine-producing country grows at least a little of this popular variety.
- Because of its sharp acidity and distinctive pungent scent, **Sauvignon Blanc** is more recognizable than most white wines.
- Germany's **Riesling** is considered by many experts to be the world's finest wine grape, but it can range confusingly from dry to sweet.
- **Pinot Gris** is capable of making ripe, rich wines but is much more popular in a lighter, brighter form as **Pinot Grigio**.
- Despite the many styles of **Moscato**, almost all are sweet and smell strongly of flowers.
- **Cabernet Sauvignon** is the most famous of Bordeaux's red varieties. The world's most expensive and age-worthy reds tend to be made with this variety.
- Although **Merlot** is better known for bargain wines, it can make powerful and elegant examples when it is not exploited for volume.
- **Pinot Noir** is Cabernet Sauvignon's closest rival for the red wine crown—indeed, many feel it makes superior wines.
- **Syrah**, also known as **Shiraz**, makes intensely flavored wines that have the **tannic structure** necessary for long-term aging.
- **Grenache** can make hefty, fleshy, bold reds that belie their pale color. They often feature **savory scents** such as white pepper or even cured ham.
- Outside of **Europe**, where regions usually grow their own **native** varieties, wine grapes of **French heritage** dominate **New World** vineyard regions.

MUST-KNOW
WINE
REGIONS

THE WORLD'S TOP VINEYARD ZONES Wine is made anywhere grapes grow, but most world-class wine is produced in a short list of regions. The classic European wine zones share certain grape-friendly qualities—dry, sun-drenched summers and mild winters—that offer great winemaking potential. There, local native grapes abound, and not all wines are made from famous varieties. In the Americas and the southern hemisphere, vines thrive in diverse landscapes—from sunny South Africa to chilly Canada. Wine can be like vicarious travel, transporting drinkers to exotic locales with the pull of a cork.

THE WINE REGIONS OF EUROPE

The complex overlapping geography of the world's wine appellations can only be truly captured in atlas form, but familiarity with the world's most significant sources is all most wine drinkers need. Europe leads global wine production, and its appellation maps are particularly intricate due to both its long wine history and each nation's complicated wine regulations.

OLD WORLD DOMINATION

For centuries, Western Europe made the world's best wines, and it still makes the most in volume. France, Italy, and Spain are the world's top three producers, despite their small size relative to many New World nations. Four more European countries make and export enough wine to be of importance in the global wine market: mainly Germany and Portugal, but also Greece and Austria. The proliferation of native grape varieties in these European nations can confound wine drinkers, as can labeling regulations that are organized differently than they are elsewhere, with the appellation playing a much more important role. Since many European wines are named for their region of origin—like Champagne and Chianti—as opposed to their grape, it's very helpful to get acquainted with the geography of the Old World.

EUROPEAN LEADERS

The majority of the world's wine is made on its original home turf in Europe. Nine European Union (EU) nations rank among the world's top 20 wine-producing countries. Among them, these seven play major roles in the global wine market: France, Italy, Spain, Germany, and Portugal, plus to a lesser extent Austria and Greece. Per capita wine consumption remains highest here as well, but it is on the decline. Since global demand for wine is growing, emphasis is shifting to international export markets.

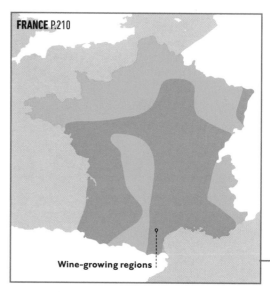

FRANCE P.210

Wine-growing regions

▲ **FIRST IN FAME, SECOND IN VOLUME**
France leads in name recognition, with its top wines such as Champagne, Bordeaux, and Burgundy commanding very high prices. It is the world's second-largest wine producer by volume.

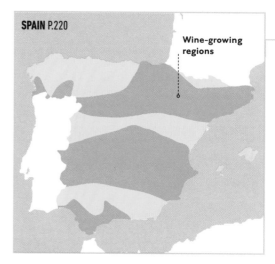

SPAIN P.220

Wine-growing regions

▲ **MOST LAND UNDER VINE**
Spain comes a close third in total production but leads in vineyard area. Many vintners are just beginning to send wine outside Europe, and exports have doubled since the turn of the 21st century.

GERMANY P.222

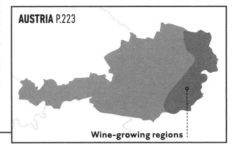

Wine-growing regions

◄ **DOUBLE TAKE**
Germany is the world's 10th-largest wine producer, making mostly white wines. However, as a nation, it consumes more than twice as much wine as it makes.

► **EASTERN BLOC**
Austria's mountains in the west are not suitable for grapes, so wine is made only in its easternmost third. It ranks 17th among the world's winemaking nations.

AUSTRIA P.223

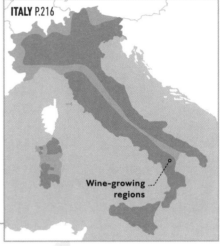

Wine-growing regions

ITALY P.216

Wine-growing regions

▲ **NUMBER ONE**
Italy is the global leader in total wine production, with the majority made from its own native grapes. Although it is smaller in area than the state of California, every one of its 20 regions makes wine.

PORTUGAL P.224

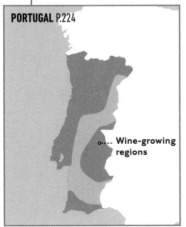

Wine-growing regions

◄ **MORE THAN PORT**
Portugal is the world's 11th-largest wine producer, despite its tiny size. It is best known for its fortified wines, but the world is now also discovering its other offerings.

► **GRAPE POTENTIAL**
Most arable terrain in Greece is vine-friendly, and grapes are grown here for wine, raisins, and currants. This nation is 18th in global wine production.

GREECE P.225

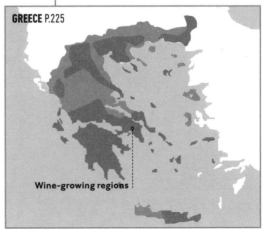

Wine-growing regions

FRANCE: BURGUNDY

The Burgundy region of central France, known as Bourgogne in French, is famous for its sensual, expressive wines and world-class native grapes. Burgundy wines are made from single varieties without blending: Chardonnay for whites, and Pinot Noir for all reds except those of the Beaujolais district. Top Burgundies are matured in new oak barrels and are renowned for their food-friendly qualities and cool-climate characteristics.

BACKGROUND

Modern fine wine traces its roots to medieval Burgundy, and the region remains notoriously difficult for wine drinkers to master.

Intricacies of geology, history, and law have resulted in 100 sub-appellations, mystifying labels, and vineyards co-owned by dozens of growers. Yet huge demand for top Burgundy drives prices sky-high, and these are some of the world's most expensive wines. Burgundy's appellation hierarchy, where wines from the smallest defined regions have the highest quality potential, serves as the model for all European wine law. Most regions stop carving out smaller appellations at the village level to avoid confusion, but individual vineyards are recognized in Burgundy. Superior vineyards may be ranked as either *premier cru* or *grand cru*, which mean roughly first class and top class respectively.

BURGUNDY'S WINE DISTRICTS

- **Bourgogne** Entry-level Burgundy wines under this name can be grown anywhere in the region. They may be described as *blanc* for crisp, dry Chardonnay whites or *rouge* for pale, earthy Pinot Noir reds.

- **Chablis** The coolest of Burgundy's districts grows only Chardonnay and is known for its tart, dry white wines that are generally unoaked.

- **Côte d'Or** The finest Burgundies come from famous villages like Chassagne-Montrachet and Chambolle-Musigny on this sunny escarpment south of Dijon.

- **Côte Chalonnaise** This lesser-known region makes small amounts of well-crafted, reasonably priced white and red wines under village appellations like Rully and Mercurey.

- **Mâconnais** This large and productive region grows mostly Chardonnay—from affordable, refreshing Mâcon, to more opulent Pouilly-Fuissé.

- **Beaujolais** Only Gamay grapes are grown here—not Pinot Noir—to make light Beaujolais reds that are fruity, affordable, and often unoaked.

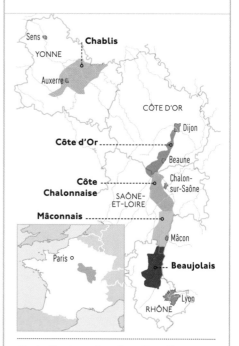

BURGUNDY AT A GLANCE

Cool-climate region of northern France

MOST POPULAR WINES
Bourgogne: basic regional whites and reds
Mâcon, Chablis: crisp, clean whites
Mercurey, Santenay: tart, earthy reds
Beaujolais: light, fruity reds

MOST PRESTIGIOUS WINES
Meursault, Puligny-Montrachet: toasty, rich whites
Vosne-Romanée, Gevrey-Chambertin: silky, earthy reds

FRANCE: CHAMPAGNE

The Champagne region produces the finest sparkling wines in the world, deftly balancing opulence and austerity. Most are very dry wines carrying the label term *brut* to indicate their lack of sweetness, and all feature refreshingly low alcohol and high acidity. The Chardonnay and Pinot Noir grapes of neighboring Burgundy are grown here, along with Pinot Meunier, a local Champenois grape in the same family that ripens much earlier.

BACKGROUND

Located near Paris, Champagne has a very cold climate compared to other wine regions, and its bubbly wines are a creative adaptation to adverse weather conditions.

Grapes are picked well below standard ripeness and used to make a still, white base wine that is very light, dry, and sour. Sugar is added to spark a second fermentation, and the resulting natural carbonation is trapped in each bottle. The wines are carefully aged on their yeasty sediment for months or years to acquire toasty flavor. Most Champagnes are blends of red and white grapes, but the grape skins are discarded early so as to make white wines. Since grape flavor varies widely from year to year, most wines are blended from multiple vintages to maintain consistency.

CHAMPAGNE'S SPARKLING WINE STYLES

- **Non-Vintage Champagne** Entry-level Champagnes are blended from multiple vintages and multiple grapes and age at least 15 months before release.

- **Vintage Champagne** Naming a year on the label indicates a superior "luxury *cuvée*" made with the fruit of a single harvest. Such wines must be aged longer on the lees—at least 3 years—before release.

- **Rosé Champagne** Pink Champagnes are usually dry and delicious, typically tinted with a splash of red Pinot Noir at the final stage of winemaking to deepen their color and flavor.

REGULATED CHAMPAGNE LABEL TERMS

- ***Brut*, *Extra-Dry*, and *Demi-Sec*** These are regulated sweetness terms used for wines that are, respectively, very dry, faintly sweet, and very sweet.

- ***Blanc de Blancs*** Meaning "white wine from white grapes," this Chardonnay-only category is a premium wine with exceptional aging potential.

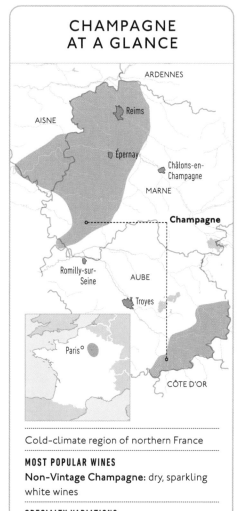

CHAMPAGNE AT A GLANCE

ARDENNES

AISNE

Reims

Épernay

Châlons-en-Champagne

MARNE

Champagne

Romilly-sur-Seine

AUBE

Troyes

Paris

CÔTE D'OR

Cold-climate region of northern France

MOST POPULAR WINES
Non-Vintage Champagne: dry, sparkling white wines

SPECIALTY VARIATIONS
Rosé: dry, sparkling pink wines
Vintage: intense premium sparkling wines
Blanc de Blancs: age-worthy sparkling white wines

FRANCE: BORDEAUX

Bordeaux is the largest wine region of France and makes the world's most influential red wine. The appellation takes its name from the port of Bordeaux, historical capital of Aquitaine. Most Bordeaux wines are red, and the region's tradition of aging the finest in new oak barrels has been adopted worldwide. Many other quality-oriented practices debuted here, too, such as estate-bottling and awarding special *grand cru* status to the finest wines estates.

BACKGROUND

Blending multiple grapes is the norm in Bordeaux, and the finest wines all spend time in new oak barrels.

Bordeaux wines have traditionally been made from a shortlist of 15 local grapes, but in practice the region's dry wines are based on one of three that have international celebrity status: Cabernet Sauvignon, Merlot, or Sauvignon Blanc. Prime sites are dedicated to Cabernet Sauvignon, which performs best along the Gironde estuary's Left Bank. Merlot, being easier to ripen, is Bordeaux's workhorse grape, and it takes the lead in value wines. However, it can also produce exceptional wines on the river's Right Bank. White Bordeaux wines are less common. Most are dry Sauvignon Blanc–based blends, but the golden Semillon-based dessert wines of Sauternes are a lusciously sweet exception. As a hedge against climate change, Bordeaux authorities recently gave provisional status to six additional grapes, but these remain experimental rarities.

BORDEAUX'S WINE DISTRICTS

- **Bordeaux** This entry-level regional appellation covers dry red, white, and rosé wines grown anywhere in the region—typically Merlot-based reds and Sauvignon Blanc–based whites. Entre-Deux-Mers whites and reds from Bourg and Blaye are of similar style.

- **Médoc** This district on the river's Left Bank is the only region where Cabernet Sauvignon thrives and dominates the blend in reds, especially in villages such as Margaux, St-Estèphe, Pauillac, and St-Julien.

- **Graves** This Left Bank region makes both red and white wines. Reds typically feature nearly equal parts Merlot and Cabernet Sauvignon.

- **Right Bank** Several small appellations on the other side of the river are known for making exceptional Merlot-based red wines, like Pomerol and St-Émilion.

- **Sauternes and Barsac** These sweet wines from the Graves region made with shriveled Semillon grapes set the world standard for luscious, oaky dessert wines.

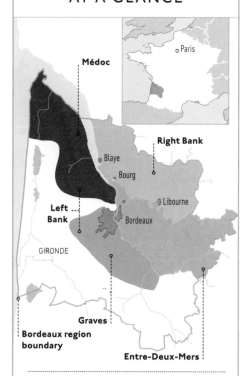

BORDEAUX AT A GLANCE

Médoc

Paris

Right Bank

Blaye

Bourg

Libourne

Left Bank

Bordeaux

GIRONDE

Graves

Bordeaux region boundary

Entre-Deux-Mers

Moderate-climate region of western France

MOST POPULAR WINES
Bordeaux, Graves: dry, mid-weight reds and whites
Médoc: superior dry reds

MOST PRESTIGIOUS WINES
Pauillac, Margaux: intense, tannic reds
Pomerol, St-Émilion: softer, fleshier reds
Sauternes: heavy, sweet, oaky whites

FRANCE: LOIRE VALLEY

The longest river in France is the Loire, whose banks are lined with cool-climate wine regions along its path westward to the sea. Ripening grapes can be a challenge in this northern region, so light white and sparkling wines are its claim to fame.

BACKGROUND

The Loire's three climate zones each favor different grapes.

The Loire's top appellations are Sancerre and Pouilly-Fumé. These make dry, unoaked white wines using 100 percent Sauvignon Blanc, which is the most famous of the Loire's indigenous grapes.

Downstream, the river's warmer mid-section is perfect for Chenin Blanc, another local variety whose white wines like Vouvray can range from bone-dry to dessert-sweet. Red and pink wines are grown here as well, the best using Cabernet Franc, a grape native to the Loire but far better known in Bordeaux. Finally, nearest the sea, Mélon grapes make brisk, dry white Muscadet.

LOIRE VALLEY'S TOP WINE STYLES

- **Sancerre and Pouilly-Fumé** Tart, dry unoaked white wines that scour the palate clean, these are the archetypes for most modern Sauvignon Blanc.

- **Vouvray and Montlouis** These Chenin Blanc whites can be dry or sweet, still or sparkling, but the most popular are lightly sweet *demi-sec* wines.

- **Chinon and Bourgueil** These pale, herbal-scented reds made with Cabernet Franc are very light, very tart, and very dry.

- **Savennières** Rare Chenin Blanc white wines that are fully dry and quite complex.

- **Coteaux du Layon and Bonnezeaux** These decadent late-harvest dessert wines are made with shriveled Chenin Blanc grapes.

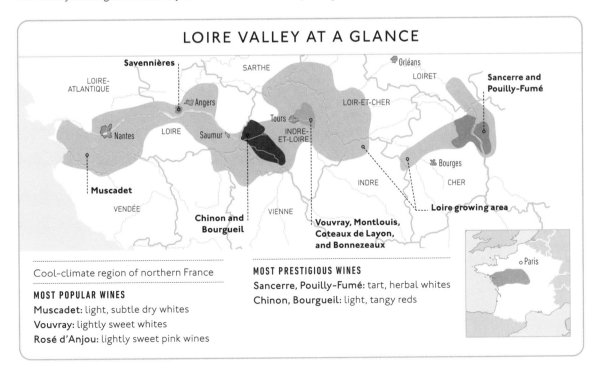

LOIRE VALLEY AT A GLANCE

Savennières · SARTHE · Orléans · LOIRET · Sancerre and Pouilly-Fumé

LOIRE-ATLANTIQUE · Angers · LOIR-ET-CHER

Nantes · LOIRE · Saumur · Tours · INDRE-ET-LOIRE · Bourges

Muscadet · CHER

VENDÉE · Chinon and Bourgueil · VIENNE · INDRE

Vouvray, Montlouis, Coteaux de Layon, and Bonnezeaux · Loire growing area

Paris

Cool-climate region of northern France

MOST POPULAR WINES
Muscadet: light, subtle dry whites
Vouvray: lightly sweet whites
Rosé d'Anjou: lightly sweet pink wines

MOST PRESTIGIOUS WINES
Sancerre, Pouilly-Fumé: tart, herbal whites
Chinon, Bourgueil: light, tangy reds

FRANCE: RHÔNE VALLEY

The Rhône Valley is a southern French region known for hearty reds and dry rosés. Rhône wines aren't as high-profile as those from Bordeaux or Burgundy, but New World vintners have embraced and popularized Rhône grapes like Syrah and Grenache (aka Shiraz and Garnacha) because they thrive in hot, dry climates. Rhône wines share a spiced flavor profile, but there are significant differences between those of the north and the south.

BACKGROUND

Below Lyon, northern Rhône vineyards cling to stony slopes as the river carves a path down from the Alps; but when it nears the Mediterranean, the valley's terrain flattens and grows wide.

The vast majority of vineyards are on the southern Rhône's warm, stony flats where it's possible to make affordable wines in volume. High-yielding Grenache, a grape of Spanish origin, dominates here, but many others of French origin—like Syrah and Mourvèdre—are blended in to boost color and flavor. The northern Rhône makes only tiny amounts of wine, because its steep vineyards are only worth cultivating where stellar wine can be made. Syrah is the only red grape planted in northern Rhône appellations.

SOUTHERN RHÔNE VALLEY APPELLATIONS

- **Côtes du Rhône** This basic appellation is known for its affordable, flavorful Grenache-based red blends from anywhere in the Rhône region, but also includes dry whites and rosés. Wines from superior townships are labeled Côtes du Rhône-Villages.

RHÔNE VALLEY AT A GLANCE

LOIRE
Vienne
Condrieu
ISÈRE
Côte-Rôtie

o Paris

Hermitage
Bourg-de-Péage
Cornas
Valence
ARDÈCHE
Côtes du Rhône

DRÔME

Montélimar

Warm-climate region of southern France

MOST POPULAR WINES
Côtes du Rhône Rouge:
flavorful, spicy reds
Côtes du Rhône Rosé:
dry pink wines

Orange
GARD
Châteauneuf-du-Pape
Avignon
VAUCLUSE

MOST PRESTIGIOUS WINES
Châteauneuf-du-Pape:
powerful, spicy reds
Hermitage, Côte-Rôtie:
intense, peppery reds

- **Châteauneuf-du-Pape and Gigondas** These prestige village appellations include the region's finest Grenache-based blends: heavy reds with intense, spicy aromatics.

- **Tavel** This appellation makes only dry Grenache-based rosés, some of the world's top pink wines.

- **Muscat de Beaumes-de-Venise** These are very sweet, very fragrant Vin Doux Naturel dessert wines.

NORTHERN RHÔNE VALLEY APPELLATIONS

- **Hermitage, Côte-Rôtie, and Cornas** These prestige wines are the world's original Syrah wines—intense, inky wines with fiercely peppery flavor.

- **Crozes-Hermitage and St-Joseph** These lesser appellations make lighter, more affordable Syrah-based wines, often more tart and earthy.

- **Condrieu** This white wine appellation makes tiny amounts of fleshy, floral, oaky Viognier wine.

FRANCE: ALSACE

Alsace is a picturesque region on France's eastern border with Germany. Its territory has long been contested, and Alsace wines reflect both French and German cultural influences quite clearly. This virtual Garden of Eden for white-wine makers combines exceptional amounts of daytime sun to develop aromatic flavor with bracingly cool nights to retain refreshing acidity, but it is too far north to make satisfying reds.

BACKGROUND

A strong German influence is immediately apparent in Alsace wines thanks to their tall, Rhine-style "flute" bottles and labels that specify grape varieties.

Fragrant grapes of Germanic origin such as Riesling and Gewurztraminer are among the region's distinctive specialties. However, French grapes—such as Pinot Gris, Pinot Blanc, and Pinot Noir—make up the majority of vineyard plantings. Both cultures shape the wine styles. Alsace wines are drier and stronger than their German counterparts, in a food-flattering style inspired by French whites like Chablis and Sancerre. While the Alsace tradition has been to make drier whites, many modern wines feature a little more overt sweetness.

ALSACE'S TOP WINE STYLES

- **Pinot Blanc and Pinot Auxerrois** Either grape can be called Pinot Blanc in Alsace, and they are often blended. These easygoing, mid-weight whites are dry and reminiscent of unoaked Chardonnay.

- **Riesling** Germany's noble Riesling is made as a stronger, drier style in France, producing piercingly fragrant whites that are food-oriented and age brilliantly.

- **Pinot Gris** This pale variant of Pinot Noir makes opulent, peachy wines in Alsace that are heavier, more aromatic, and often sweeter than those made the Italian way under the name Pinot Grigio.

- **Gewurztraminer** This grape of Austrian heritage is famous for its Moscato-like floral perfume. It makes unusually dense and flavorful white wines that may be fully dry or lightly sweet.

- **Pinot Noir** Very few Alsace wines are red, and only Burgundy's elegant, earthy Pinot Noir grape may be used.

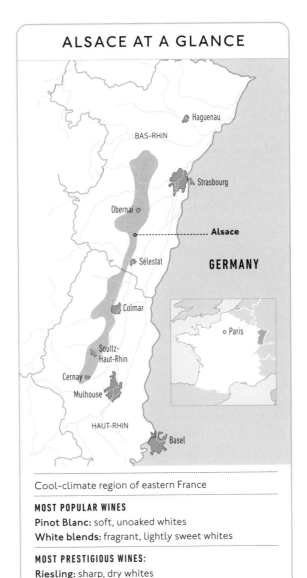

ALSACE AT A GLANCE

Cool-climate region of eastern France

MOST POPULAR WINES
Pinot Blanc: soft, unoaked whites
White blends: fragrant, lightly sweet whites

MOST PRESTIGIOUS WINES:
Riesling: sharp, dry whites
Pinot Gris: opulent, peachy whites

ITALY: TUSCANY

Toscana, or Tuscany, is a region of rolling hills on Italy's Mediterranean coast. A broad swathe of arable land in a mountainous nation, it is the top wine-producing region in Italy, which is itself the world's top wine producer in volume. The past half-century has brought fine-wine sensibilities to a region previously more focused on volume.

BACKGROUND

Most Tuscan wine is red and almost all are blends based on the region's native Sangiovese grape.

Tuscany's best-known wine is Chianti, from its largest appellation. This central zone has many subdistricts, of which Chianti Classico—in the heartland of Chianti, between Florence and Siena—is most renowned. Farther south, the hilltop town of Montalcino is famed for its unique thick-skinned variant of Sangiovese, the Brunello clone that makes deeper, darker, more age-worthy wines.

TUSCANY'S TOP WINE APPELLATIONS

- **Chianti and Chianti Classico** These medium-bodied, dry red wines are made primarily with Sangiovese grapes. Pale, tart, and tannic, these food-friendly offerings are Italy's best-known red wines.
- **Brunello di Montalcino and Rosso di Montalcino** Made entirely with a special clone of Sangiovese, these Montalcino reds have more color, flavor, and body. Brunellos are aged premium wines; Rossos are younger.
- **Toscana Rosso** This is a broad, loosely regulated category that can range from simple everyday reds to luxurious prestige wines. They are usually, but not always, Sangiovese-based.
- **Vernaccia di San Gimignano** These brisk, unoaked white wines are dry, tart, and lightweight.

TUSCAN TERMS

- **Riserva** This is a legal term for wines that have been given more aging before release.
- **Super-Tuscan** This informal term was originally coined for nontraditional prestige wines, often using French grapes, but has become a catchall description for Tuscan blends.

TUSCANY AT A GLANCE

Chianti Classico
Chianti
Pisa
Livorno
Florence
Siena
Perugia
TUSCANY
UMBRIA
Montalcino
Terni
Bolgheri
Toscana region boundary
LAZIO
Rome

Moderate-climate region of central Italy

MOST POPULAR WINES:
Chianti: tart, dry Sangiovese-based red blends
Toscana Rosso: Chianti-like, but more modern—often riper and fruitier

MOST PRESTIGIOUS WINES:
Brunello di Montalcino: intense, aged reds made with a superior type of Sangiovese grapes (Brunello clone)
Bolgheri: strong reds, often blends of Cabernet Sauvignon and Merlot with Sangiovese; this is the home of the original Super-Tuscans

ITALY: PIEDMONT

Piedmont has Italy's longest history of fine winemaking, thanks in part to its proximity to France, Europe's pioneer in valuing quality over quantity. The region is framed by mountains to the north, west, and south, and its name means "foothills." Frequent fogs and cloudy days can make it difficult to ripen grapes here, so hillside vineyards with extra sun exposure are essential in this challenging environment.

BACKGROUND

Chilly Piedmont's most popular wine is sweet, frothy Asti.

Along with its semi-sparkling cousin Moscato d'Asti, it is the international benchmark for fragrant, lip-smacking Moscato wines. However, Piedmont's claim to fame in the fine-wine world is a matched pair of epic red wines: Barolo and Barbaresco. They are named for adjacent towns and made entirely with Nebbiolo grapes. Only prime sunward slopes allow this stubborn grape to ripen properly, so two other local red grapes, Barbera and Dolcetto, occupy lesser sites. Both once made only simple, young wines of everyday quality, but top Barberas have come to rival the great Nebbiolos. Less well known but well worth seeking out are Piedmont's dry whites: lean, sharp Gavi and soft, fragrant Arneis.

PIEDMONT'S RED WINE STYLES

- **Barolo and Barbaresco** These premium, age-worthy tannic red wines are made with Nebbiolo grapes. Austere, earthy, and very powerful, they rank among Italy's most respected.
- **Barbera d'Alba and Barbera d'Asti** Medium-bodied red wines made with Barbera grapes, these range from seafood-friendly refreshers to denser, oaky prestige wines. All have high levels of acidity.
- **Dolcetto** These medium-bodied red wines made with Dolcetto grapes are often fruity and fresh tasting, with mild acidity and a vivid purple color.

PIEDMONT'S WHITE WINE STYLES

- **Asti and Moscato d'Asti** These are sparkling and semi-sparkling sweet wines from Moscato Bianco grapes. Fermentation is interrupted to make a style that is essentially half sparkling wine, half white grape juice.
- **Gavi** Subtle, dry whites made with Cortese grapes.
- **Arneis** Fragrant, dry whites made with Arneis grapes.

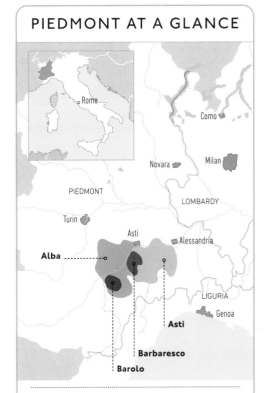

PIEDMONT AT A GLANCE

Moderate-climate region of northwestern Italy

MOST POPULAR WINES:
Asti: sweet, sparkling Moscato
Barbera: tart, mid-weight reds
Dolcetto: soft, fruity, young reds

MOST PRESTIGIOUS WINES:
Barolo: tannic, full-bodied Nebbiolo reds that require aging
Barbaresco: Barolo-like but a touch lighter and brighter

ITALY: TRIVENETO

Known collectively to locals as the Triveneto, the northeastern Italian regions of the Veneto, Friuli-Venezia Giulia, and Trentino-Alto Adige have a distinct cultural heritage that is reflected in their wines. This trio surrounds Venice, stretching from fertile coastal plains in the south, to alpine valleys in the north. Most of Italy's vines of French heritage are planted in the Triveneto, many for centuries, and wine labels usually name their grape alongside their appellation in the manner of Germany and the New World.

BACKGROUND

The Triveneto's climate favors white wines, and the region's most visible exports—Pinot Grigio and Prosecco—are both light and fresh tasting.

Low-lying areas make quirky whites, like floral Friulano and nutty Soave from Garganega grapes; however, French varieties such as Pinot Grigio, Chardonnay, and Pinot Bianco dominate in the mountains of Alto Adige and Trentino. The Triveneto's most popular reds are the fruity Valpolicellas of Verona, but the region also makes excellent mid-weight reds from native grapes like Lagrein and Refosco, as well as from French imports like Merlot and Refosco.

NORTHEASTERN ITALY'S TOP WHITE STYLES

- **Pinot Grigio** The Venezie and cooler regions of Trentino and Alto Adige are famous for their light, mild Pinot Grigios, but warmer vineyards in Friuli often make stronger, oakier variations on the style.

- **Prosecco** This light, off-dry sparkling wine hails from the Veneto foothills north of Venice but can be made in most of the Triveneto.

- **Friulano** This fragrant grape's lively dry wines were long known as Tocai but were forced to change names to respect the legacy of Hungary's Tokaj region.

NORTHEASTERN ITALY'S TOP RED STYLES

- **Valpolicella** Verona's environs produce delightful blends based on the Corvina grape, ranging from easy-drinking value wines to denser premium bottlings.

- **Amarone della Valpolicella** This luxury variant of Valpolicella is made by drying grapes for a month or more before winemaking—one of Italy's strongest wines.

- **Merlot and Pinot Noir** Northern Italy is better known for whites but also makes many light, refreshing, cool-climate reds that are typically labeled by grape.

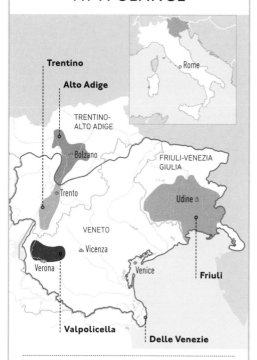

THE TRIVENETO AT A GLANCE

Trentino

Alto Adige

Rome

TRENTINO-ALTO ADIGE

Bolzano

FRIULI-VENEZIA GIULIA

Trento

Udine

VENETO

Vicenza

Verona

Venice

Friuli

Valpolicella

Delle Venezie

Cool-climate region of northeastern Italy

MOST POPULAR WINES
Pinot Grigio delle Venezie: light, unoaked whites
Prosecco: light sparkling wines
Valpolicella: light, fruity reds

MOST PRESTIGIOUS WINES
Amarone: intense, raisiny reds

ITALY: THE SOUTH

Southern Italy has historically been a source of cheap, anonymous reds, but radical changes have seen it emerge as a quality wine zone with great potential. Centuries of poverty shaped this region's wine culture, favoring quantity over quality as a rule. For most of the 20th century, roughly 90 percent of the region's wine was sold off in bulk, and when a regulatory structure was imposed, its requirements were initially manipulated to prop up the mediocre status quo.

BACKGROUND

Most wines from mountainous southern Italy name their grape on the label.

Naples provides the area with a metropolitan anchor, but the culture and economy remain rooted in farming and fishing. Most southern Italian wines are red and share a flavor profile that suits the local cuisine, regardless of their grape: tart, dry reds, with fiery spiced-fruit flavors. However, the region's rare white wines are turning heads, too, for their uncommon aromatic character.

SOUTHERN ITALY'S TOP RED GRAPES

- **Aglianico** This ancient grape produces pungent wines that combine power and grace in appellations like Campania's Taurasi and Aglianico delle Vulture in Basilicata.

- **Montepulciano** This productive grape from Abruzzi is best known for easy-drinking mid-weight reds, but stronger premium bottlings show great promise.

- **Negroamaro** The "black and bitter" grape of Puglia is known for its dark color and high tannins, which provide the Salento region's premium wines with great aging potential.

- **Nero d'Avola** Sicily's most aromatic red variety is often sold under its own name, delivering Syrah-like reds of uncommon intensity.

- **Primitivo** Known as Zinfandel in California, this Croatian grape is widely planted in Puglia, making lighter, tangier, and more seafood-friendly wines.

SOUTHERN ITALY'S TOP WHITE GRAPES

- **Fiano** This grape from Campania offers dry white wines that feature Rhône-like weight and floral aromatics.

- **Insolia** This Sicilian native makes snappy, tart wines in cool mountain vineyards that will please fans of Sauvignon Blanc.

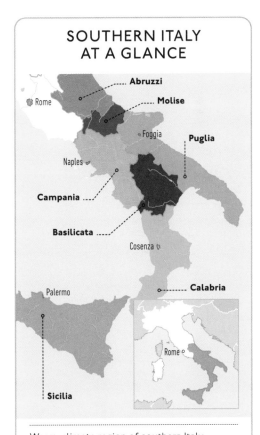

SOUTHERN ITALY AT A GLANCE

Rome
Abruzzi
Molise
Foggia
Puglia
Naples
Campania
Basilicata
Cosenza
Palermo
Calabria
Rome
Sicilia

Warm-climate region of southern Italy

MOST POPULAR WINES
Montepulciano: soft, fruity reds
Nero d'Avola: tart, aromatic reds
Primitivo: tangy, earthy reds

MOST PRESTIGIOUS WINES
Taurasi: intense, spicy reds
Fiano: rich, fragrant whites

SPAIN

Spain has a longer history of growing grapes and making wine than most of France, but the region was slower to modernize and focus on quality due to its unique history. In a remarkably swift turnaround sparked by joining what is now the European Union (EU), Spain is making world-class wines, many of which blend Old World traditions with New World techniques to broaden their international appeal. Modest Spanish wines deliver exceptional value in both reds and whites, while ritzier appellations like Priorat and Ribera del Duero rival France's Burgundy and Italy's Barolo for critical acclaim.

BACKGROUND

From hot, sunny Andalusia to cool, breezy Galicia, Spain is blessed with varied, vine-friendly terrain and noble native grapes.

Red Tempranillo and white Albariño earn rave reviews for wines like Rioja and Ribera del Duero, Toro and Rías Baixas, in the north. Farther east, indigenous reds like Garnacha and Monastrell carpet Mediterranean shores from Valencia to Barcelona. In challenging climates, Spain makes quirky wines that are wholly unique, like fizzy, underripe Basque Txakoli and nutty, sun-dried Sherries from Jerez. Elsewhere, Spanish vintners follow international inspiration—applying the methods of French Champagne to local Catalan grapes to make sparkling Cava, for example.

SPAIN'S TOP RED WINE STYLES

- **Rioja** Spain's best-known red wines are mid-weight Tempranillo blends from this Atlantic-cooled zone south of Bilbao, often given long barrel aging.
- **Ribera del Duero and Toro** Tempranillo is also the main grape in

warmer Castilla y León, making denser, stronger wines in these appellations along the banks of the Duero River.

- **Priorat and Montsant** These zones near Barcelona make blended reds with old-vine Garnacha and Cariñena. Priorats have more power and prestige, while Montsants are more affordable.
- **Tempranillo, Garnacha, and Monastrell** These three Spanish grapes are often encountered in value-oriented wines that are labeled by grape, particularly those from central Spain.

SPAIN'S TOP WHITE WINE STYLES

- **Cava** A wildly popular sparkling wine from Catalonia's Penedés region, Cava is typically made with Spanish white grapes but bottle-fermented and lees-aged in the style of French Champagne.
- **Albariño** This bracing, seafood-friendly white from the Galician coastal zone of Rías Baixas combines the attributes of refreshing Pinot Grigio and aristocratic Chablis.
- **Jerez-Xérès-Sherry** The world's most diverse range of fortified

wines are made from white Andalusian grapes and range from crisp, bone-dry Manzanilla to dark, raisiny Moscatel.

QUALITY AND MATURITY

Spanish vintners have traditionally judged wines by their intensity and aged the best wines longer in casks before they are sold. This is common practice in the red wine regions of Europe, but it has been regulated in Spain. Vintners making ambitious wines can earn the right to use three honorific label terms of ascending rank—*crianza, reserva,* and *gran reserva*—to help their customers recognize wines of superior quality that have spent extra time maturing in oak barrels and bottles. Many of these wines are aged beyond their legal mandates and released when they near their peak, often 5–10 years after their harvest for the top wines.

SPAIN AT A GLANCE

Txakoli
Bilbao

Albariño/Rías Baixas

Priorat and Montsant

Bierzo

Valdeorras

Ribera del Duero

Navarra

Toro

Rioja

Rueda

Barcelona

Madrid

Cava/Penedés

Utiel-
Requena

Jumilla

Seville

Castilla-La Mancha

Jerez-Xérès-Sherry

KEY
- GREEN SPAIN
- NORTH CENTRAL SPAIN
- SOUTHERN SPAIN
- MEDITERRANEAN SPAIN

Mixed-climate region
of Western Europe

SPANISH RED WINES
Rioja: tangy, oaky reds
Ribera del Duero: dense,
oaky reds
Garnacha: strong, young reds

SPANISH WHITE WINES
Cava: toasty, sparkling whites
Albariño: light, fragrant whites
Sherry/Jerez: nutty, fortified
whites

GREEN SPAIN

Spain's coldest wine region is on its
Atlantic coast. Green Spain's best
wines are dry, tart, low-alcohol
whites that are rarely aged or oaked.
The most renowned are Galicia's
Albariños from Rías Baixas, which
have exceptional finesse and
aromatics. Less common but with
a cult following are the fizzy, cidery
Txakoli wines of the Basque Country
and fragrant Godello wines of
Valdeorras. Red grapes don't ripen
well here, aside from the Mencía
grape of Bierzo.

NORTH CENTRAL SPAIN

Spain's finest reds are almost all
Tempranillo-based wines from
regions north of Madrid. Rioja is the
source of Spain's most iconic reds and
the first Spanish region to embrace
barrel aging. Nearby Navarra makes
Spain's top dry rosés just north of
the Garnacha grape's native terrain in
Aragón. South and west across the
mountains, the Castilian plateau ripens
Tempranillo more fully in Ribera del
Duero and Toro to make denser, darker
reds, while Rueda makes snappy
Verdejo-based whites.

SOUTHERN SPAIN

Most premium wines here are fortified
whites from coastal Andalusia. The
most successful are the Sherries of
Jerez—from pale, bone-dry Manzanilla
and Fino, to dark, sticky-sweet Pedro
Ximenez and Moscatel. Castilla-La
Mancha, south of Madrid, makes most
of Spain's value wines.

MEDITERRANEAN SPAIN

The milder Mediterranean coast's
best-known wines come from
Catalonia in the north. Cava, the
world's number one sparkling wine, is
made in the Penedés, also known for
its whites and reds labeled by grape.
Nearby Priorat and Montsant make
some of Spain's strongest reds, blends
based on ancient bush-planted
Garnacha. Further south, inky
Monastrell-based reds are made in
Murcia's Jumilla appellation and
jammy Utiel-Requena wines from
Valencia use the Bobal grape.

GERMANY

Fine-wine traditions reach back to the Middle Ages in Germany, and until a century ago, German white wines were considered the world's finest. A harsh climate led generations of vintners to focus single-mindedly on perfecting wines from the cold-hardy Riesling grape. Modern Germans like their Rieslings dry these days, but export markets prefer the lightly sweet, low-alcohol style traditionally associated with German wines.

BACKGROUND

Germany's original quality wine regions, the Mosel and Rheingau, are now joined by the up-and-coming Pfalz, Nahe, and Rheinhessen.

The country's cold climate yields tart, low-sugar grapes that don't always ripen fully enough to make balanced dry wines. Since sun-derived grape sweetness and flavor are prized, Germany developed its unique *Prädikat* labeling system around these properties. Top wines are ranked according to the ripeness achieved by their grapes, measured in terms of sugar content at harvest (see below). Degrees of ripeness can vary widely—not only by vineyard location but from one vintage to the next.

GERMANY'S TOP WHITE WINE REGIONS

- **Mosel** The coldest of Germany's wine regions makes some of its most ethereal wines. Most are sweet-tart, and many fall below 10% alcohol.
- **Rheingau** This warmer zone allows the possibility of greater ripeness. Many wines are classically light and sweet-tart or fully dessert-sweet, but some modern styles are heavier in weight and bone-dry.

GERMAN LABEL TERMS

- *Kabinett, Spätlese,* and *Auslese* Indicating *Prädikat* wines, or regulated quality wines, of modest, superior, and exceptional ripeness respectively, these terms are usually found on wines that range from lightly sweet to fully sweet. They may also appear on dry wines, since they technically assess grape sweetness, or potential alcohol, not wine sweetness after fermentation.
- *Beerenauslese and Eiswein* These are sweet *Prädikat* wines made from late-harvest grapes of higher and higher sugar content, the final stage being freeze-concentrated icewine from grapes picked in midwinter.
- *Trocken and halbtrocken* Meaning dry and half-dry respectively, these terms indicate wine sweetness and are used for wines with very little perceptible sugar.

GERMANY AT A GLANCE

Cold-climate region of Northern Europe

MOST PRESTIGIOUS WINES
Mosel Riesling: lightly sweet, appley whites
Rheingau Riesling: denser, peachier whites

OTHER GERMAN STYLES
Spätburgunder: light, Pinot Noir reds
Müller-Thurgau: sappy, young whites

AUSTRIA

Austria is a German-speaking country, but its wines have their own distinctive identity, driven by unique geography and reliance on native grape varieties. Most Austrian wine is grown in Lower Austria (Niederösterreich) along its eastern borders with Hungary and Slovakia. More white than red is made in Austria's cool, continental climate, but all grapes ripen more easily than in Germany, yielding wines that are stronger and more often fermented dry.

BACKGROUND

Most Austrian wines are named for their grape, with Grüner Veltliner being the dominant variety.

Its wines appeal most to fans of Sauvignon Blanc and dry Riesling, since it has similar leafy scents and tangy acidity, but in top sites it can produce wines with the richness of great Chardonnay. Austria's top dark grapes, Zweigelt and Blaufränkisch, make quirky, mid-weight reds. But the most decadent of Austria's offerings are its luxurious dessert wines. These opulent, honeyed wines follow a German-style ripeness-based *Prädikat* system that indicates increasing degrees of sweetness and concentrations, from lightly sweet *Spätlese* on up.

AUSTRIA'S TOP WHITE WINE STYLES

- **Grüner Veltliner** This grape is rarely seen outside Austria. Its unoaked white wines have an herbal scent and can range from fruity quaffers to complex age-worthy beauties that can rival white Burgundy's complexity.

- **Riesling** While Riesling is not Austria's most planted grape, it makes some of the country's most interesting fine wines, often dry and mid-weight in a style similar to that of Alsace.

- **Weissburgunder and Grauburgunder** These are German aliases for two French grapes better known by their Italian names—Pinot Bianco and Pinot Grigio—that make mild, dry white wines in Austria.

AUSTRIA'S OTHER WINE STYLES

- **Zweigelt and Blaufränkisch** These cold-hardy reds are close relations that make surprisingly robust and flavorful wines with moderate alcohol.

- *Eiswein and Ausbruch* With milder weather than Germany, Austria uses similar grapes and techniques to make hyper-sweet dessert wines and specializes in winter-harvested icewines.

AUSTRIA AT A GLANCE

Vienna

CZECH REPUBLIC

Niederösterreich

Wien

Vienna

Bratislava

Burgenland

HUNGARY

Graz

Steiermark

SLOVENIA

Cool-climate region of Northern Europe

TOP WHITE WINES
Grüner Veltliner: light, herbal whites
Riesling: tangy, dry whites

OTHER WINE STYLES
Zweigelt: vibrant, tangy reds
Eiswein: sticky-sweet dessert wines

PORTUGAL

Portugal is very small, roughly half the size of Florida, but has a remarkable array of indigenous grapes. When varietal wines from well-known grapes were all the rage in the 1980s and '90s, Portugal was at a disadvantage: Its grapes were unfamiliar to the wider world and were often interplanted in the vineyard rather than blended at the winery. Luckily, Portugal's isolation helped preserve what is now seen as a treasure trove of grapevine diversity.

BACKGROUND

Historically, fortified wines were considered Portugal's only world-class offerings, and until recently, other Portuguese wines were rarely seen outside the country.

Sweet Port, made in northern Portugal's Douro Valley, had pride of place as the world's favorite dessert wine. Less well known but of similar heritage and quality, Madeira hailed from a tropical island that flies the Portuguese flag. The only other Portuguese wines with a global presence were the enjoyable lightly sweet rosés from the Minho, exemplified by mass-market brands like Mateus and Lancers. Nowadays, there is growing interest in this coastal nation's other quirky wines—from spritzy white Vinho Verde, to denser reds from Dão and Alentejo.

PORTUGAL'S CLASSIC FORTIFIED WINE STYLES

- **Porto** The strong, sweet Port wines from the Douro Valley are fortified mid-fermentation with distilled spirit. The vast majority are red blends, often of six or more native Portuguese and Spanish grapes.

- **Madeira** Portugal's other great fortified wine is named for the tropical island where it's made. These tart Sherry-like wines taste nutty, range from dry to fully sweet, and are made from both white and red grapes.

PORTUGAL'S OTHER WINE STYLES

- **Vinho Verde** These light, tart wines are affordable, fizzy, and refreshing. They may be white, pink, or red, but they are called *verde* ("green") because their grapes are picked underripe.

- **Alentejo/Tejo** These warmer inland regions in the south make riper, stronger wines—typically dry reds labeled by grape variety.

- **Douro** These dry red wines from Port country are made with the same grapes as Port, yielding vibrant flavor and deep color.

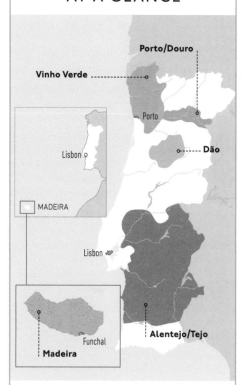

PORTUGAL AT A GLANCE

Porto/Douro

Vinho Verde

Porto

Lisbon

Dão

MADEIRA

Lisbon

Funchal

Madeira

Alentejo/Tejo

Mixed-climate region of Western Europe

TOP STANDARD WINES
Vinho Verde: light, fizzy whites
Alentejo: deep, tangy reds

TOP FORTIFIED WINES
Porto: sweet, liqueur-like reds
Madeira: sweet-tart, nutty whites

GREECE

Ancient Greece played an important role in the spread of vines and the refinement of winemaking across the Mediterranean, but the region's wine culture stagnated under centuries of rule by the Ottoman Turks. It wasn't until the country joined the European Economic Community in 1981 that significant changes took place and premium Greek wine became a modern reality.

BACKGROUND

Most of the coastal terrain of Greece shares a sunny Mediterranean climate and, like Italy, is home to hundreds of indigenous grape varieties.

While most of these are cultivated as table grapes or for drying into raisins or currants, a few dozen have earned a reputation for making excellent wine, such as white Assyrtiko from the southern islands or red Agiorgitiko from the mainland north. In numerical terms, a few large-scale producers dominate Greek production, and export markets still see mostly traditional styles like piney *retsina* or sweet Muscat and Mavrodaphne. But increasingly, small vintners are introducing the wine world to new flavors from Greece, most of which are dry, tangy wines well suited to the region's traditional Mediterranean cuisine.

GREECE'S TOP DRY WINE STYLES

- **Assyrtiko** The tart, dry whites of Santorini are based on this hard-scrabble vine whose wines please fans of French Sauvignon Blanc.

- **Moschofilero** This fragrant pink grape is cultivated on the Peloponnese peninsula, where it makes plump whites that feature exotic floral aromas.

- **Xynomavro** Found planted throughout central and northern Greece, this tannic grape has a name meaning "sour red." It performs best in Naoussa and Amynteo.

- **Agiorgitiko** This grape makes the tangy, mid-weight reds of Nemea and is sometimes compared to Italian Sangiovese for its versatility and food-friendliness.

GREECE'S TOP SWEET WINE STYLES

- **Muscat and Mavrodaphne** Sweet, white Muscat and red Mavrodaphne abound in Greece. They are typically fortified wines made using the Port method.

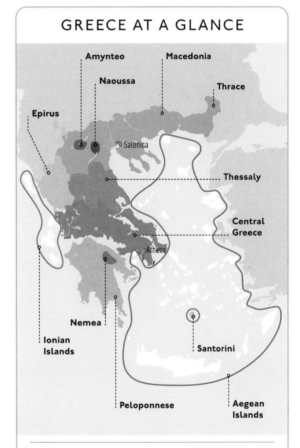

GREECE AT A GLANCE

Amynteo
Macedonia
Naoussa
Thrace
Epirus
Salonica
Thessaly
Central Greece
Athens
Nemea
Ionian Islands
Santorini
Peloponnese
Aegean Islands

Warm-climate region of Mediterranean Europe

UP-AND-COMING WHITE WINES
Santorini: bracing, tart whites
Moschofilero: aromatic, floral whites

UP-AND-COMING RED WINES
Xynomavro: tannic, dry reds
Nemea: tangy, mid-weight reds

THE WINE REGIONS OUTSIDE EUROPE

New World wine areas occupy vine-friendly zones across the western and southern hemispheres. Learning the wine geography of these regions is less essential than it is for Europe, since their appellations are generally fewer in number, larger in area, and far less complicated in regulatory terms. Wines are also typically labeled by grape variety in the New World, and fewer different grapes are grown, which makes label navigation easier for wine drinkers. However, this can make it harder to differentiate between individual wines and can limit the diversity of flavors available.

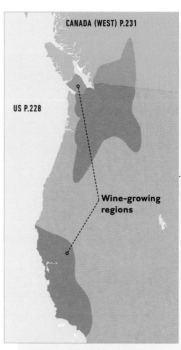

CANADA (WEST) P.231

US P.228

Wine-growing regions

NEW WORLD LEADER
The United States makes the most wine outside Europe, ranking fourth in global terms. Nearly 90 percent of this is produced in sunny California.

SMALL BUT SELF-SUFFICIENT
In volume, Canada ranks only 28th in global wine production, but it produces nearly half of its own national consumption.

CANADA (EAST) P.231

Wine-growing regions

EXPORT FOCUS
Chile's vineyard area is smaller than that of neighboring Argentina, but it exports twice as much wine. In terms of production volume, it is sixth in the world.

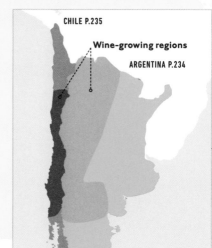

CHILE P.235

Wine-growing regions

ARGENTINA P.234

SECOND OUTSIDE EUROPE
Argentina is the New World's second-largest wine producer and number five in global terms. More than 70 percent of its vineyards are in the province of Mendoza.

BRAVE NEW WORLD

While the former colonies are often larger in territory than European nations, they tend to make less wine. The United States leads the pack as the world's fourth-largest producer, but over 85 percent of its output never leaves its shores. Argentina, Chile, Australia, and South Africa are also among the global top 10, and all have a growing presence in export markets. New Zealand generates far less volume but looms larger in international importance than many bigger rivals. Canada's total is even smaller, and little is exported, but it gains in visibility by making world-class icewines and by being a part of the lucrative North American market for fine wine.

TOP NEWCOMERS

These seven New World nations rank among the world's top wine regions due to their importance in the global market. By volume, China, Russia, and Brazil each make more wine than New Zealand or Canada, but their international visibility is very low in comparison.

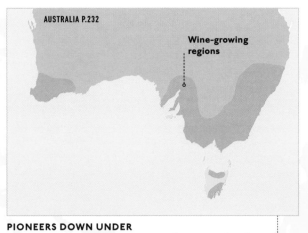

AUSTRALIA P.232

Wine-growing regions

PIONEERS DOWN UNDER
Australia is the seventh-largest wine producer on earth and a leading center of research and innovation in the New World. Its vineyard regions hug its cool southern coast.

A TRUE NEWCOMER
New Zealand was barely a blip on the global wine radar until its Marlborough Sauvignon Blancs rose to prominence in the 1990s. Today, it ranks 16th in world wine production.

NEW ZEALAND P.236

Wine-growing regions

SOUTH AFRICA P.237

Wine-growing regions

GRAPES OF THE CAPE
Very little wine is made on the African continent, but South Africa's Western Cape region has an ideal climate for growing grapes. The nation is the world's ninth-biggest source of wine.

USA: CALIFORNIA

California makes considerably more wine than any New World wine nation. Only France, Italy, and Spain produce a greater volume than the United States, and nearly 90 percent of US wine comes from California. Ample sunshine and scarce rainfall result in exceptionally ripe grapes whose bold flavor and rich texture have become the hallmark of California wine.

BACKGROUND

Wines from California have shown promise since the mid-1800s, but major setbacks derailed their progress for a hundred years.

An insect plague at the turn of the 20th century ravaged vineyards, and then came Prohibition, which outlawed winemaking from 1920 to 1933. Bulk wines were the first to recover in the postwar period, and it wasn't until the 1970s that fine wine reemerged in earnest. California vintners and American wine drinkers are now leading forces in the global wine boom that has revolutionized the wine world. Today, the state makes some of the finest wines on earth.

California's broad Central Valley produces most of its everyday wine but is too hot for fine wines. Premium wine zones hug Northern California's coastline in two groups, above and below the Bay Area, where fruit's rush to ripeness is slowed by the cool Pacific air. The North Coast is known as California's wine country, encompassing the most famous and historic wine appellations, such as Napa Valley and Sonoma's Russian River Valley. The Central Coast stretches south from San Francisco to Santa Barbara, incorporating up-and-coming wine zones like Paso Robles and Monterey.

CALIFORNIA'S TOP RED WINE STYLES

- **Cabernet Sauvignon and blends** This grape revels in California's sunshine, making world-class wines in sheltered zones like Napa Valley and Paso Robles.

- **Zinfandel** This uniquely American style is a warm-climate specialty, high in color and alcohol, with flavors like baked berry desserts, especially in areas with older vineyards, such as Lodi and Dry Creek.

- **Merlot** In places like Napa Valley, this underrated grape excels, making plush reds that are high in flavor but low in harsh tannin.

- **Pinot Noir** This thin-skinned grape thrives in cool coastal regions like Sonoma, Santa Barbara, and Monterey, making seductive, mid-weight wines.

CALIFORNIA'S TOP WHITE WINE STYLES

- **Chardonnay** California's number one white grape produces plump, fruity wines particularly well in cool zones such as Sonoma and Santa Barbara. They may or may not feature oaky flavor.

- **Sauvignon Blanc** Sometimes labeled Fumé Blanc, this grape's wines are often less acidic and less herbal here than in other countries.

NEW WORLD NAMES

All wine labels list their appellation, but in the US and most New World nations the grape variety is shown as the primary style indicator. Pioneering vintners in new territories studied the European classics to decide which vines to plant and how to make wine, but the Old World system of naming wines by place could not be adopted. Since this practice is based on historical precedent, it is simply not workable in emerging regions. Instead, American wine laws are designed for maximum flexibility for vintners, with basic controls on the accuracy of label statements. Unlike European wine regulations, they do not attempt to impose quality standards for each appellation by limiting crop yields, or to require regional consistency by mandating specific grapes be planted.

CALIFORNIA AT A GLANCE

Sonoma County

Napa Valley

San Francisco Bay Area

Sacramento

Monterey

Paso Robles

Santa Barbara County

Fresno

Bakersfield

Los Angeles

Central Valley

Washington, D.C.

KEY
NORTH COAST
CENTRAL COAST

Warm-climate region of North America

TOP WHITE WINES
Chardonnay: rich, toasty whites
Sauvignon Blanc: brisk, tangy whites

TOP RED WINES
Cabernet Sauvignon: dense, powerful reds
Zinfandel: strong, jammy reds
Pinot Noir: elegant, mid-weight reds

NORTH COAST

This is California's prestige wine zone, consisting mainly of four counties north of the San Francisco Bay: Sonoma and Mendocino on the coast, and Napa and Lake counties one step inland.

Sonoma is the largest North Coast county and features the most variation in climate. Foggy zones near the water, such as Russian River Valley and Carneros, specialize in cool-climate wine styles like Chardonnay, Pinot Noir, and sparkling wines. Inland appellations like Alexander Valley and Dry Creek Valley are warmer and better known for heavier reds such as Cabernet Sauvignon and Zinfandel.

Napa Valley is a fraction of Sonoma's size but is denser in vineyards and more famed for its premium wines. Sheltered by a range of mountains, Napa is warmer, with a climate more uniformly suited to red grapes. The Bordeaux partners Cabernet Sauvignon and Merlot perform exceptionally well and are often blended, both in flat valley-floor appellations like Rutherford and in steeper terrain like Howell Mountain and Stags Leap District.

Mendocino County's climate and landscape are similar to Sonoma's, but its vineyards are more sparse.

Lake County's climate resembles that of the northern Napa Valley.

CENTRAL COAST

Most often seen as a regional appellation on wine labels, the Central Coast stretches south from the Bay Area to Santa Barbara. Despite being farther south than the North Coast, these valleys are typically cooler, because coastal ridges channel breezes from the Pacific deeper inland. Santa Barbara, San Luis Obispo, Monterey, and Santa Cruz counties make exceptional Chardonnay, Pinot Noir, and Syrah, while the enclave of Paso Robles is better known for bolder reds such as Cabernet Sauvignon and Zinfandel.

USA: PACIFIC NORTHWEST

California may dominate American production, but its neighbors in the Pacific Northwest also make exceptional wines. Washington and Oregon share a border and are similar in size but couldn't be more different in their wine styles. Oregon is known for small amounts of ultra-premium Pinot Noir and Pinot Gris, while Washington produces far greater amounts of more affordable Merlot, Cabernet Sauvignon, Chardonnay, and Riesling.

BACKGROUND

The stylistic divide reflects the region's geography, with east and west separated by the Cascade Mountains.

Oregon's primary wine region, the Willamette Valley south of hipster Portland, is wedged between the Cascades and the coast. Being relatively cool, it is perfect for growing Pinot Noir, a fickle grape that's hard to farm but of great value when perfected.

Washington's main wine region is the Columbia Valley near the cow towns of Yakima and Walla Walla, on the eastern "dry side" of the Cascades, where irrigation makes it possible to grow fruit on an arid plateau. Daytime sun and warmth here can easily ripen thick-skinned red grapes, while cold desert nights allow white grapes to retain brisk acidity.

PACIFIC NORTHWEST'S TOP RED STYLES

- **Oregon Pinot Noir** The Willamette Valley is considered one of the world's best sources of soulful, expressive Pinot Noir outside Burgundy.

- **Washington Merlot** This variety makes uncommonly powerful wines in the Columbia Valley, belying the grape's "soft and fruity" reputation.

- **Washington Syrah** Exceptional full-bodied Syrahs are made in central Washington, with deep color and strong aromatic character.

PACIFIC NORTHWEST'S TOP WHITE STYLES

- **Oregon Pinot Gris** These light-hearted wines are mid-weight, unoaked, and dry. Most are heavier than Italian Pinot Grigio but milder than French Pinot Gris.

- **Washington Riesling** Washington makes lovely Chardonnay and Sauvignon Blanc, but its Riesling is something truly special, usually in the lightly sweet German style.

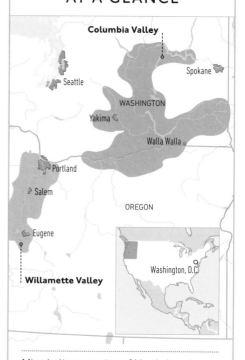

PACIFIC NORTHWEST AT A GLANCE

Mixed-climate region of North America

MOST POPULAR WINES
Washington Riesling: lightly sweet whites
Oregon Pinot Gris: dry, unoaked whites

MOST PRESTIGIOUS WINES
Oregon Pinot Noir: silky, mid-weight reds
Washington Syrah: intense, peppery reds

CANADA

Visitors to Canada are often surprised to discover a thriving wine industry: vineyards don't fit our picture of the great white north. Very little local wine is exported, but Canada supplies about half of its own wine needs. The region's most prestigious style is its luscious icewine, inspired by the *Eiswein* of Germany and Austria. These dessert wines are made with overripe grapes whose juice is freeze-concentrated by being left on the vine well into winter.

BACKGROUND

Most vines can't survive more than a day or two below 14°F (–10°C), so only a handful of Canada's southernmost regions have mild enough winters to support grape growing.

Canadian fine-wine making began in earnest in Ontario's Niagara Peninsula, between the Great Lakes. But today there is just as much to be excited about on the west coast in British Columbia, where the Okanagan Valley is making exceptional wines in the foothills of the Canadian Rockies. While both regions are known for cool-climate specialties, such as Riesling and Pinot Noir, the west is more consistently able to ripen grapes that need more sun and warmth, such as Cabernet Sauvignon and Syrah.

CANADA'S TOP WINE STYLES

- **Icewine** The most prized Canadian wines are freeze-concentrated dessert wines made by letting grapes hang on the vine into January.
- **Riesling** This cool-climate white grape from Germany adapts well to Canadian vineyards and is typically made in a light, sweet-tart German-inspired style.
- **Cabernet Franc** This thinner-skinned relation of Cabernet Sauvignon thrives in Canada, making snappy, cedary, mid-weight red wines.
- **Pinot Gris** Performing brilliantly in these sunny but cool regions, Pinot Gris here is more often modeled on rich Alsace Pinot Gris than lighter Italian Pinot Grigio.
- **Pinot Noir** This is a finicky variety whose lighter reds can be heart-breakers, but Canadian vintners are showing promise in both Ontario and British Columbia.
- **Syrah** Though this variety is more commonly seen from warmer climates, British Columbia is making great Syrah: intense, fragrant reds.

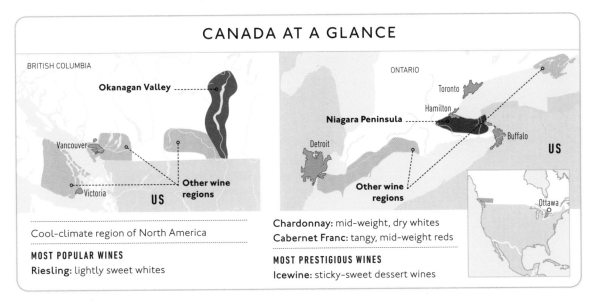

CANADA AT A GLANCE

BRITISH COLUMBIA

Okanagan Valley

Vancouver

Victoria

Other wine regions

US

ONTARIO

Toronto

Hamilton

Niagara Peninsula

Detroit

Buffalo

US

Other wine regions

Ottawa

Cool-climate region of North America

MOST POPULAR WINES
Riesling: lightly sweet whites

Chardonnay: mid-weight, dry whites
Cabernet Franc: tangy, mid-weight reds

MOST PRESTIGIOUS WINES
Icewine: sticky-sweet dessert wines

AUSTRALIA

Most of Australia's landmass is too hot for vineyards, but many of its southern regions are blessed with a grape-friendly Mediterranean climate. Many 20th-century innovations in grape growing and winemaking were pioneered here. These ideas have been spread by Australian "flying winemakers" who consult on northern-hemisphere harvests while their own vines lie dormant in winter.

BACKGROUND

In global wine production, Australia ranks among the top ten, but because it is sparsely populated, its vintners focus more on exports than those of most New World nations.

As a rule, the finest wines are grown in areas that benefit from the cooling influence of either the mountains or the sea, with most falling within 185 miles (300 km) of the coastline between Sydney and Adelaide. Everyday bargain wines are more often grown in warmer, irrigated zones farther inland, such as the flats of the Murray River's drainage basin.

Many grapes of European heritage are grown in Australia, including Chardonnay and Cabernet Sauvignon. However, the most widely planted is Australia's signature grape Shiraz. This French variety, known elsewhere as Syrah, adapts well to warm climates. Australian wines are almost always labeled by grape variety, according to the New World norm, with one significant difference: Australian blends must list all of their component grapes in order of importance.

GEOGRAPHICAL INDICATIONS

Many of Australia's most popular wines are labeled under a single generic appellation—South East Australia—that encompasses the vineyards of five states and about 90 percent of the country's plantings. Premium wines, though, are often bottled under the names of smaller appellations, or geographical indications, following the "smaller is better" model that dominates the wine industry.

AUSTRALIA AT A GLANCE

Warm-climate region of the southern hemisphere

TOP WHITE WINES
Chardonnay: rich, fruity whites
Riesling: tart, dry whites

TOP RED WINES
Shiraz: dark, jammy reds
Grenache: strong, raisiny reds

WESTERN AUSTRALIA

Perth

Margaret River

Other wine regions

WESTERN AUSTRALIA

Western Australia is a massive state that covers the western third of the continent. Only a small area of coastal land near Perth is suitable for wine growing, and this remote region is home to more small independent vintners than larger winery conglomerates. The finest wines come from the Margaret River region, famed for its outstanding Chardonnay and Shiraz.

SOUTH AUSTRALIA

Many of Australia's most respected appellations lie within a short drive of Adelaide. South Australia's southeastern corner offers prestige wine regions along the coast and volume-oriented vineyards in the drainage of the Murray River. Barossa and McLaren Vale loom large as the country's most famous sources of Shiraz, while the cooler Clare Valley makes stunning dry Riesling. To the south, Coonawarra and the Limestone Coast produce some of Australia's top Cabernet Sauvignon and Chardonnay.

AUSTRALIA'S TOP RED WINE STYLES

- **Shiraz** The Australian moniker for the French Syrah grape, it makes strong, dark, flavorful wines, ranging from fun and fruity to brooding and intense.

- **Grenache blends** Other Rhône varieties also thrive in Australia, like Grenache, often seen in so-called GSM blends with small amounts of Shiraz and Mourvèdre in the model of French Côtes du Rhône.

- **Cabernet Sauvignon and blends** Cabernet Sauvignon is less widely planted here than in other New World zones and makes lighter, tangier wines. It is often blended with Shiraz.

AUSTRALIA'S TOP WHITE WINE STYLES

- **Chardonnay** Australia's diverse geography provides a wide array of Chardonnays—from tangy cool-climate, to luscious warm-climate styles—many of which are unwooded, or made without new oak flavor.

- **Riesling** Australian Riesling wines are most often dry and tart in the French Alsace style, with exceptional lime and green-apple aromatics.

- **Semillon** Best known for making France's top sweet wine, Sauternes, this lesser-known grape makes all sorts of whites down under—from bone-dry to sticky sweet, from brisk and unoaked to toasty and barrel-fermented.

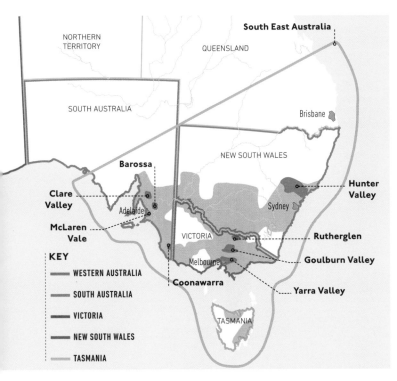

KEY
— WESTERN AUSTRALIA
— SOUTH AUSTRALIA
— VICTORIA
— NEW SOUTH WALES
— TASMANIA

NEW SOUTH WALES

The Hunter Valley, near Sydney, was the first region to show fine-wine potential, so Australia's wine history is deepest in New South Wales. This area makes outstanding Chardonnay and Shiraz and has a unique take on Semillon. Most New South Wales production comes from the interior's irrigated Murray–Darling basin, but ambitious vintners also make exciting wines in the cooler hills of the Great Dividing Range, in smaller appellations like Mudgee and Hilltops.

TASMANIA

The smallest and coldest of Australia's states is Tasmania, an island in the frigid Southern Ocean. The region is known for white wines such as Chardonnay and Riesling, but it is earning recognition for great sparkling wine and Pinot Noir as well.

DOWN UNDERDOG

Most New World regions make Sauvignon Blanc their main dry, unoaked white wine, but Riesling unexpectedly fills this role in Australia, making some of the most thrilling dry wines from this underrated grape.

VICTORIA

The southernmost and coolest of Australia's mainland states, Victoria specializes in cool-climate wine styles, like white and sparkling wines. The coastal zones that surround Melbourne, such as Yarra Valley and the Mornington Peninsula, are ideal for grapes that love cool weather, like Pinot Noir and Chardonnay. The mountain valleys of the Great Dividing Range, like Goulburn, make great dry whites and reds in a variety of styles. Northeastern towns like Rutherglen are famed for their sweet and fortified dessert wines, or "stickies," modeled on Port and Sherry.

ARGENTINA

The largest wine region in Latin America is Argentina's Mendoza province. At the edge of the Andes, the high-elevation plains of Cuyo produce dense, lush red wines that have long been prized in South America, but only in the past 30 years has the rest of the world discovered why. Mendoza's high-desert climate provides the daytime warmth and sunlight that are crucial for ripening dark-skinned grapes. At night, temperatures plunge, slowing the process and retaining the acidity necessary for making balanced fine wines.

BACKGROUND

Compared to its competitors, Argentina was slow to export its wines beyond South America.

Since its primary grape varieties were unfamiliar to international wine drinkers—just blips on the global wine radar—they faced an uphill struggle to earn confidence from wine drinkers. Argentina's number one grape, Malbec, has no significant presence outside Argentina. Since it doesn't ripen well in its native France, it is rarely grown there or elsewhere. White Torrontés is a local cultivar descended from European varieties but with no Old World roots of its own. In Argentina, both grapes perform brilliantly, producing wines with enough personality and complexity to turn heads around the world.

ARGENTINA'S TOP WINE STYLES

- **Malbec** These flavorful red wines are dark and rich, with floral and earthy aromatics. Premium bottlings are dense and oaky, while bargain brands tend to be lighter and fresher.

- **Bonarda** This milder red variety is of no generic relation to the Bonarda grown in northern Italy. It makes soft, fruity wines in an easy-drinking style.

- **Torrontés** The floral-scented white wines from this variety are typically dry and unoaked but have a distinctive scent reminiscent of Moscato.

- **International varieties** Argentina's climate is well suited to grape growing, and recent years have seen more experimentation with famous grapes such as Chardonnay and Cabernet Sauvignon.

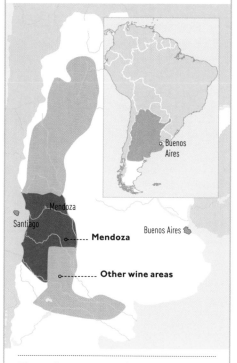

ARGENTINA AT A GLANCE

Mendoza

Santiago

Buenos Aires

o------ **Mendoza**

o---------- **Other wine areas**

Warm-climate region of the southern hemisphere

POPULAR WHITE WINES
Torrontés: fragrant, dry whites

POPULAR RED WINES
Malbec: dark, intense reds
Bonarda: lighter, fruity reds

CHILE

This long, thin nation has made wine since the Spaniards arrived, but only recently has it earned a reputation for world-class quality. Chile stretches for thousands of miles, sandwiched between the Pacific Ocean and the peaks of the Andes. Its wine country occupies a series of temperate vine-friendly valleys surrounding the capital city of Santiago that enjoy months of cloud-free skies in summer and mild winters.

BACKGROUND

French vine varieties were introduced to Chile in the 19th century, and today Cabernet Sauvignon, Merlot, and Sauvignon Blanc dominate Chile's exports, along with Chardonnay.

Chile also cultivates one grape found nowhere else: Carmenère. Prized for its intense flavor, this red Bordeaux grape had nearly disappeared from France by the late 1800s, but was discovered a century later to be thriving in Chile, interplanted with and mistaken for its close relation Merlot.

With exceptional sun exposure, cool nights, and a long, dry growing season, Chilean vineyards rarely experience the wet weather and insect pests that can plague other regions. Chile's geography, along with low-cost land and labor, attracted international capital and winemaking expertise in the 1990s, largely through joint ventures, and wine quality has since improved by leaps and bounds.

CHILE'S TOP WINE STYLES

- **Carmenère** Chile's signature grape makes red wines that resemble its Bordeaux relations Merlot and Cabernet Sauvignon, producing intense dark, herbal-scented wines with great quality potential.
- **Cabernet Sauvignon, Merlot, and Bordeaux-style blends** Chile makes exceptional wines with Bordeaux varieties, and many of the best are blends from regions like the Maipo, Rapel, and Aconcagua valleys. Compared to other New World examples, they are often made with more French-style food-oriented sensibilities.
- **Chardonnay and Sauvignon Blanc** White grapes thrive in regions of Chile that are close to the cool Pacific coast, such as the Casablanca Valley. Chardonnays tend to be crisp and refreshing, with premium examples displaying the most oak character. Sauvignon Blanc is also proving itself here, making snappy, tart wines that echo the unoaked, herbal New Zealand style.

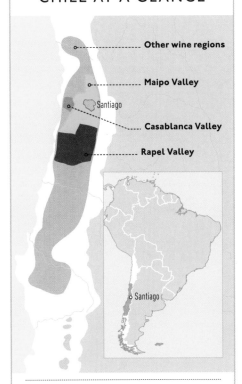

CHILE AT A GLANCE

Other wine regions

Maipo Valley

Santiago

Casablanca Valley

Rapel Valley

Santiago

Mixed-climate region of the southern hemisphere

MOST POPULAR WINES
Merlot: mid-weight, dry reds
Chardonnay: mid-weight, dry whites

MOST PRESTIGIOUS WINES
Carmenère: dark, aromatic reds
Cabernet Sauvignon: intense, age-worthy reds

NEW ZEALAND

While it may seem like a major player in the wine world today, known primarily for its refreshing whites, New Zealand's wine industry is much younger than those of its New World competitors. Significant commercial plantings began in the 1970s as part of a national economic restructuring, and a decade later one style emerged to vault New Zealand to international stardom: Sauvignon Blanc from Marlborough. Tart, dry, and citrusy, these wines follow the Loire Valley model of Sancerre but feature more exuberant herbal and tropical aromatics.

BACKGROUND

New Zealand is unusual among New World regions in that its vineyard zones are colder in climate—more like cool, moist European zones than the hotter, drier areas that make wine in the United States, Australia, and South America.

However, this is exactly what makes New Zealand wines stand out, combining the clean, fruity character associated with modern New World winemaking with the bracing acidity and food-friendliness more commonly found in Old World European wines. The dramatic success of New Zealand Sauvignon Blanc has boosted global interest in other cool-climate growing regions and the fortunes of tart, dry white wines fermented in stainless steel everywhere.

NEW ZEALAND'S TOP WHITE WINE STYLES

- **Sauvignon Blanc** Unoaked and bracingly tart, these mid-weight, dry whites feature intense citrus and green-herb aromas. The distinctive Marlborough style has become an international benchmark for the grape.

- **Chardonnay** This grape performs brilliantly in New Zealand's coastal zones, especially along the North Island's southeastern coast. Its wines feature a zing of green-apple acidity and may be oaked or unoaked.

NEW ZEALAND'S TOP RED WINE STYLES

- **Pinot Noir** New Zealand is one of very few international regions where this fickle grape makes world-class wines, particularly in the South Island's Central Otago region.

NEW ZEALAND AT A GLANCE

[Map of New Zealand showing: Auckland, NORTH ISLAND, Wellington, Marlborough, Hawke's Bay, Christchurch, SOUTH ISLAND, Central Otago]

Cool-climate region of the southern hemisphere

POPULAR WHITE WINES
Sauvignon Blanc: tart, unoaked whites
Chardonnay: tangy, dry whites

POPULAR RED WINES
Pinot Noir: pale, tangy reds

SOUTH AFRICA

Most of the African continent is too hot for making fine wine, but South Africa's Western Cape region has a Mediterranean climate. This area was one of the first in the New World to explore fine-wine making in the late 1600s and was considered a top-notch source by the early 1800s. However, the 20th century proved a major setback for South African wine. After insect pests ravaged vineyards early on, vintners turned to bulk wines, and for decades much of the country's wine was distilled into cheap brandy.

BACKGROUND

Before Apartheid was lifted in the 1990s, trade embargoes blocked many export markets just as the global wine boom began.

Now, however, the world is once again recognizing the distinctive character and quality potential of South African wines, which have an uncommon complexity. Volume-oriented production is based in sun-baked interior valleys, while premium winemaking occurs in cooler coastal zones. Despite the warmth, South Africa's wines are often more food-oriented and less fruit-forward than other New World regions.

SOUTH AFRICA'S TOP RED WINE STYLES

- **Pinotage** South Africa's signature grape variety was a result of crossing Pinot Noir with an obscure but productive Rhône grape called Cinsaut. Pinotage makes intense red wines, with a smoky, meaty scent.

- **Cabernet Sauvignon and blends** South African Cabernet-based wines have an earthy character, particularly from Stellenbosch and Paarl.

SOUTH AFRICA'S TOP WHITE WINE STYLES

- **Chenin Blanc** This Loire Valley grape thrives in South Africa, making thrilling styles—from light and sweet to full-bodied, dry, and barrel-fermented.

- **Chardonnay and Sauvignon Blanc** These grapes do well in cooler zones, often in a crisp, dry style.

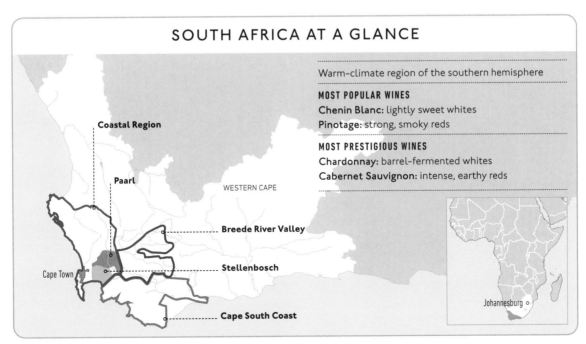

SOUTH AFRICA AT A GLANCE

Coastal Region

Paarl

WESTERN CAPE

Breede River Valley

Stellenbosch

Cape Town

Cape South Coast

Johannesburg

Warm-climate region of the southern hemisphere

MOST POPULAR WINES
Chenin Blanc: lightly sweet whites
Pinotage: strong, smoky reds

MOST PRESTIGIOUS WINES
Chardonnay: barrel-fermented whites
Cabernet Sauvignon: intense, earthy reds

HOW TO ACHIEVE VINLIGHTENMENT

Many wine lovers find themselves overwhelmed when they try to learn more about wine. The key is to focus on core concepts, not reams of details, and you must also learn to trust your senses.

TRUST YOUR TASTE BUDS

- **Embrace your personal tastes** Don't feel pressured to adopt "sophisticated" preferences. The only person worth impressing is yourself.

- **Keep an open mind** Wine perceptions change with food context, temperature, and even with your moods, so give wine styles second chances.

- **Bend the rules** Don't let stodgy ideas about wine cramp your style. Wine for breakfast? Wine on the rocks? Mixing two wines? Go for it!

- **Try new things** You'll never know what's out there until you taste it, so don't let past preferences limit your future experiences.

- **Don't sweat the details** It's easy to get wrapped up in wine minutiae on multiple levels, but this can interfere with your enjoyment.

LEARN HOW TO DESCRIBE WINE

- **Use simple language** Wine's most important traits can be summed up in a few basic terms that have concrete meanings.

- **One sense at a time** Each of our senses except hearing contribute to our perceptions of wine.

- **Assess sensory traits** Rank color and color depth, sweetness and acidity, fruit and oak flavors, and weight and tannin from low to high.

- **Don't worry about naming smells** Identifying flavors and scents takes practice. Stick to judging their overall intensity and broad categories.

- **Notice highs and lows** Most wines fall in the crowded middle ground on most traits, but qualities that fall outside the norm are the most distinctive and most helpful.

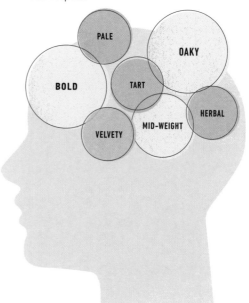

GET COMFORTABLE WITH WINE SHOPPING

- **Judge books by their cover** Packaging decisions often provide useful insights into a vintner's philosophy and intended audience.

- **Crunch the numbers** When shopping, a few helpful figures—such as a wine's age, price, and alcohol content—can tell you a fair amount about its style.

- **Read the fine print** Getting familiar with two main labeling formats can make wine shopping less confusing.

- **Stick to your budget** Drinking well doesn't require overspending. Consider wines off the beaten track or bargains in 3-liter boxes.

- **Ask for help without ceding control** Sales staff can be a great resource, but there's no need to let them decide what you'll spend.

EXPLORE THE FULL RANGE OF WINE STYLES

- **Sample by the glass when possible** Restaurants offer a great opportunity to expand your experience without committing to a full bottle.

- **Step outside your comfort zone** Most people start out with a favorite wine style, but there may be others you'd like even more if you tried them.

- **Don't judge on first sip alone** Wine's acidity can shock the palate at first. Second and third sips give a better sense of a wine's attributes.

- **Take a moment to savor new wines** Give them a minute or two of your attention, then file them in your mental database for future reference.

- **Use your words** Putting your thoughts into words will make it much easier to remember a wine later, even if they aren't wine terms or spoken aloud.

TASTE THE SUNSHINE IN YOUR GLASS

- **Work backward from alcohol content** In dry wines, alcoholic strength is a reasonably accurate indicator of ripeness, with 13.5% being the norm.

- **Look for more oomph in stronger wines** The higher a dry wine's alcohol content, the more likely it is to taste more intense and to be oak-aged.

- **Look for more refreshment in lighter wines** The lower a dry wine's alcohol content, the more likely it is to taste milder and feature bracing acidity.

- **Smell the ripeness spectrum** Herbal, earthy aromas occur most in cool-climate wines, while warm-climate offerings have more dessert-like spiced-fruit aromas.

- **Don't let sweet wines fool you** Alcohol content isn't a sound guide with specialty categories like dessert wines and fortified wines.

PLAY WITH YOUR FOOD AND WINE

- **Put wine in its place** Wines are better suited to supporting roles than being the star of the show, so let food flavors take center stage.

- **Strike a match** Try choosing wines that echo a dominant feature of the dish, whether it's taste or texture, scent or flavor.

- **Adjust for food chemistry** Pick tangy wines for salty foods and sweet wines for sugary foods.

- **Watch out for spicy heat** Don't forget that alcohol amplifies the "burn" of spicy foods, while lighter wines tame the flames.

- **Defy the rules** It's your wine, so you should drink what you like—even if it doesn't fit the usual guidelines.

TEST YOUR WINE SAVVY

- **Guess which wines aren't fully dry** Wines with sweetness are usually low-alcohol whites, and those with the most sugar often come in small bottles.

- **Deduce which wines taste of oak** Older, more expensive wines are more likely to feature the flavor of oak barrels.

- **Predict which wines are most tart** Wines that are younger, lower in alcohol, or that come from colder regions are generally most acidic.

- **Estimate which wines are most food-oriented** Many European wines can seem a little too tart or dry alone and are designed for serving with salty food.

- **Gauge which wines will have the strongest flavor** Wines from warm regions with high alcohol often feature very bold fruit and oak flavors, especially those from the Americas and southern hemisphere.

RELAX AND ENJOY YOURSELF

- **Banish snobbery and pretense** It's a grave injustice that something as inherently relaxing and social as wine is seen as uptight and standoffish.

- **Live vicariously through grapes** Wine's scent evokes its region and culture—close your eyes, take a sniff, and be transported.

- **Skip the traditional homework** Don't bother memorizing reams of wine data. If there's something you want to know, you can always look it up.

- **Live in the moment** Wine is perfectly suited for reminding us that any day spent with friends and family is an occasion worth celebrating.

- **Raise a glass to your success** You've taken on a thorny topic and picked up some tricks of the trade. Here's to a long life of eating well and drinking better!

GLOSSARY

Acidic, acidity
Detectable presence of sourness; one of the six true taste sensations detected by the tongue's taste buds.

Age-worthy
A descriptive term for wines that resist oxidation, thanks to high levels of natural preservative components like tannin or acidity.

Alcohol
Ethanol, the psychoactive component in adult beverages such as wine; an organic compound derived from sugar by the metabolism of living yeast cultures.

Antioxidant
A substance that inhibits oxidation or its properties, as with phenolic compounds in grape skins, such as tannin.

Aperitif
Alcoholic drink designed to pique the appetite before a meal. Many light wines are served as aperitifs, but the term "aperitif wines" may also refer to flavored and fortified wine-based drinks, such as all Vermouths, Lillet, or Dubonnet.

Appellation
A formal region-of-origin statement indicating where a wine's grapes were grown; mandatory on all wine labels.

Archetype
Original specimen or pattern on which subsequent examples are modeled.

Aromatics, aromatic
Wine components perceived by the nose that convey sensations of both scent and flavor; their presence in fragrant wine.

Astringency
In wine, the mouth-drying effect of tannins, which are found in grape skins and suppress salivation.

Barrel
Round storage vessels made of oak used in the maturation or fermentation of wine.

Barrel aging
A common practice in the making of red wine, where young, fresh wines are stored in oak barrels after fermentation for anywhere from a few weeks to a few years.

Barrel fermentation
A white-winemaking practice, where grape juice is fermented into wines in oak barrels that are often aged on their yeast sediments for up to a year.

Barrique
Traditional French-style 225-liter wine barrel; associated with styles where new oak flavor is imparted during aging.

Biodynamic
A form of natural farming that maintains the interrelated ecosystem, including soil, plants, and animals; also a rigorous agricultural certification system that prohibits use of nonnative and synthetic vineyard treatments and organizes cultivation around the lunar and astral cycles.

Bitter, bitterness
One of the six sensations detected by the tongue's taste buds (an example might be hops in beer); often confused with the tactile astringency of tannin in wine (as in black tea).

Blend
A wine made from several different grape varieties.

Body
Descriptive term for wine's texture, typically driven by alcohol content. See **Weight**.

Bold
Term for wine that is high in aromatic or flavor intensity.

Brand name
A wine's commercial identity; may be the vintner's name or a proprietary name for a product line.

Browning
In wine, a visible sign of age and oxidation. In food, the color change associated with caramelization and Maillard reactions in certain cooking methods, like searing, frying, and grilling.

Brut
Regulated label term for sparkling wines that have no perceptible sweetness; one step drier than extra-dry.

Bulk wine
The lowest quality category of wines, often used in cheap blends.

Carbonated
Descriptive term for wine that is bubbly, releasing carbon dioxide when opened.

Cellaring
Bottle-aging; maturing of wine by purchasers after release.

Color compounds
Phenolic wine components such as tannins that are derived from grape skins and that give color and flavor to red and rosé wines.

Complex
Wine term for simultaneous sensations in wine; typically refers to the presence of many pleasing scents and flavors, especially those generated during fermentation or aging.

Concentrated
Descriptive term for wines with higher-than-average intensity of olfactory scents and flavors; typically refers to fruit rather than oak.

Condense
In chemistry, to change physical matter from gas to liquid. The opposite of evaporate.

Cork taint, corked
Detectable spoilage of wine caused by contact with natural cork; most commonly the presence of TCA (short for 2,4,6-trichloroanisole), a compound that gives wine an unpleasant mildew smell.

Corks
Wine bottle stoppers punched from the bark of the cork oak tree.

Crianza
A regulated Spanish wine-label term for aged wines of good quality; the lowest tier of a hierarchy that includes *reserva* and *gran reserva*.

Crisp
Descriptive term for wines that have a standard, moderately tangy level of acidity.

Cru
A French wine term used for recognizing sites that produce superior wines, typically qualified as top-level *grand cru* or next-best

premier cru. Often translated as "growth," its meaning is closer to "rank" and is complicated by different criteria in different regions, enforced with varying rigor.

Cuvée, cuvée name
Wine term for a particular bottling of wine, often needed for clarification when more than one is made from the same appellation and grape variety; from the French word for vat. In some regions, a narrower usage indicates a blended wine.

Decanting
Process used for removing wine from its bottle (usually into a vessel known as a decanter) before it is served, either to separate an older red wine from its sediment or to aerate a younger wine.

Demi-sec
Regulated French label term for wines that have an overt sweetness, typically balanced sweet/tart wines.

Dessert wine
A category of wine that features overt, intense sweetness.

Distilled spirits
High-alcohol beverages such as brandy and whiskey, made from low-alcohol fermented products like wine and beer by vaporizing their alcohol and condensing it in a separate container.

Dry, dryness
Descriptive term for wine with no noticeable presence of sugar; the opposite of sweet and the norm for the majority of wines. Often confused with the mouth-drying effect of tannin in red wines.

Earthy
Term for wine smells that are reminiscent of the outdoors and the farming environment, such as mulch, stones, and fallen leaves.

Estate
A vineyard owned by the vintner who makes its wine, allowing them to farm their own grapes rather than purchasing from a grower, as is the norm.

Estate-bottled
Regulated label term used primarily for New World wines, indicating that the vintner owns and farms the vineyard from whose grapes the wine was made.

Esters
Volatile aromatic compounds that are a major source of scent and flavor in many fruits, as well as in wine.

Evaporate
In chemistry, to change physical matter from liquid to gas; vaporize. The opposite of condense.

Everyday wine
Simple wines that are fairly priced; a step above bulk wines and the cheapest of the fine wines.

Extra-dry
Regulated label term for sparkling wines that have faint perceptible sweetness; one step sweeter than *brut*.

Fermentation
Main stage of winemaking that converts juice into wine; the process by which all alcoholic drinks are made, where living yeast organisms consume and metabolize sugar, breaking it down into alcohol and carbon dioxide.

Fine wine
Premium wines, in which quality is a consideration.

Finish, length
Term for a wine's aftertaste; the duration of this desirable trait is useful for assessing wine quality.

Flabby
Descriptive term for wine that has lower-than-average levels of acidity.

Flavor
In common usage, taste sensations derived from eating and drinking; for purposes of wine analysis, smells that reach the olfactory nerves from the mouth internally via the retro-nasal passage.

Food-friendly, food-oriented
Term for wines that are designed to taste best with food, especially the acid-blocking, fruit-enhancing effects of salt in food.

Fortified wine
Category of wines ranging from 15% to 20% alcohol that contain added distilled spirits, such as Port and Sherry.

Freeze-concentrate
To reduce the water content of a liquid by freezing it and removing its solid ice crystals, as for icewine.

Fruit, fruity
In common usage, sweet edible produce, typically containing seeds of a flowering plant; in wine tasting, a descriptive term for wine smells deriving from grapes or winemaking. When it is prominent, a wine may be described as fruity, fruit-driven, or fruit forward.

Full-bodied
Term for wines of richer-than-average texture. See **Heavy**.

Generic
Common or nonexclusive. In New World, refers to wines that do not specify a grape variety on their label. In Old World, refers to the most basic wines of a region—for example, generic Chianti as opposed to Chianti Classico from a superior subdistrict.

Grand cru
A French wine quality classification. See **Cru**.

Gran Reserva
A regulated label term for top-quality aged wines in Spain and South America.

Grape variety
Cultivar of the grape species *Vitis vinifera*; includes the many grape types used in winemaking.

Green
In fruit, underripe; in wine, displaying characteristics associated with low ripeness, such as high acidity and leafy, herbal aromatics.

Grip
Descriptive term for the astringent, mouth-drying, tactile sensation of tannin found in many red wines.

Harsh
Term for a strong presence of either astringent mouth-drying tannin in red wine or unusually high alcohol in any wine.

Headspace
The upper section of a wine-glass bowl that remains empty in order to allow swirling and to concentrate wine aromas.

Heavy
Descriptive term for wines with a rich, mouth-coating texture; associated with wines that have an alcohol level of more than 14%.

Herbal
Descriptive term for wines whose aromas and flavors resemble herbs, leaves, or vegetables.

Icewine
A dessert wine made by freeze-concentration—harvesting frozen grapes in midwinter.

Indigenous
Originating from a particular region.

Jammy
Descriptive term for wines whose aromas and flavors suggest fruit that has been cooked or sweetened.

Kabinett
Regulated German label term indicating a quality wine made from grapes of standard ripeness; the lowest of the *Prädikat* levels, *Kabinett* wines are often lightly sweet and low in alcohol, but some are heavier and drier.

Lactone
Aromatic ester found in oak barrels that contributes to oaky flavor in wine.

Late harvest
Label term used for sweet wines, indicating that the fruit was left on the vine longer to become sweeter and riper.

Legs
The drips that form when wine is swirled in a glass—a visible indicator of wine's weight.

Length
See **Finish**.

Light
Term for wines with a sheer, delicate texture, associated with wines of below 13% alcohol.

Maderization
Flavor changes in wine caused by exposure to heat; named for the wines of Madeira.

Maturation
Winemaking stage where wine rests in barrels, tanks, or bottles after fermentation.

Mature
Term for wine at its peak and requiring no further aging.

Mid-weight
Descriptive term for wines with moderate texture, neither light nor heavy, associated with wines of 13–14% alcohol.

Mild
Descriptive term for wine that is low in aromatic or flavor intensity.

Mousse
Descriptive term for the carbonation of wine.

Mouth-drying
See **Tannin**.

Mouthfeel
Tactile sensations of food and drink perceived in the mouth.

Naked
See **Unoaked**.

Neutral barrel
An oak barrel that has been used to store wine for at least 3 years, reducing its ability to impart new oak flavors.

New oak
Oak barrels or products that have not previously come into contact with wine; the flavors and scents that these impart.

New World
Collective term for the wine regions of the Americas and southern hemisphere.

Oak, oaked, oaky
In common usage, a type of tree or its wood; in wine tasting, a descriptive term for wine smells that derive from its contact with new oak barrels or oak flavoring agents during winemaking. Wines exhibiting such smells are said to be oaked.

Off-dry
Lightly sweet; not fully dry.

Old World
Wine term for traditional wine regions of Europe.

Olfactory
Of or relating to the sense of smell.

Organic
Type of natural farming and agricultural certification of its products that prohibits the use of synthetic chemical treatments.

Oxidation, oxidized
The primary source of wine spoilage resulting from prolonged exposure to air, typically avoided during winemaking; a descriptive term for its effects in wine, such as

reduced freshness; browned color; and nutty, cooked-fruit smells.

Pairing
Choosing a wine for its suitability as a flattering partner for a particular food item or dish.

Palate
Technically, the soft flesh of the mouth; also used informally to refer to a person's sensitivity to tastes and smells or their wine preferences.

Phenolic compounds
Color and flavor compounds found in grape skins, such as tannin and anthocyanin; many are antioxidants with natural preservative properties.

Point scores
Numerical rankings of wines bestowed by magazines and critics; typically assessments of abstract quality on a 100-point scale.

Potential alcohol
Sugar content of grapes prior to fermentation, defining the upper limit of possible alcohol content in wines made from them.

Premier cru
A French wine quality classification. See **Cru**.

Preservative
A substance that slows spoilage and oxidation; may be naturally present in wine, as with tannin, or an additive, such as sulfur dioxide.

Proprietary name
A wine name that is particular to a specific vintner, as with a *cuvée* name or brand name.

Racy
Descriptive term for wine that has high levels of acidity.

Refreshing
Term used for wines whose acidity provides a bracing, restorative sensation.

Reserve, Reserva, Riserva
Wine-label terms that suggest superior quality; regulated in Spain, Italy, and South America, but with no legal standards elsewhere.

Rich
Descriptive term for wines that feel more thick or viscous in the mouth than average; see also **Heavy**.

Ripening, ripeness
The final stage of fruit development in the last weeks before harvest, where exposure to sunlight and warmth causes grapes to become sweet, juicy, and ready to pick.

Rosé
Category of wines that are pink in color, made by giving clear grape juice brief contact with dark grape skins during winemaking.

Saccharomyces
The genus of "sugar-eating" yeasts that produce beverage alcohol; the yeast category used for making wine, beer, and bread.

Salt, saltiness
A common food component that reduces the perceived acidity of wine when served alongside; one of the six true taste sensations detected by the tongue's taste buds.

Sediment
A solid precipitate that settles from a liquid.

Sensory
Of or relating to perceptions of sight, smell, taste, touch, and hearing.

Sharp
Descriptive term for wines that have a high "tart" level of acidity.

Single-vineyard
Wine made from grapes grown on one plot of land.

Soft
Descriptive term for wines that are low in astringent tannin; also sometimes applied to low-acid wines.

Sommelier
Wine steward or wine-specific server in a restaurant; typically also the person in charge of wine purchasing.

Sour
See **Acidic**.

Sparkling
Descriptive term for carbonated wine with bubbles.

Spätlese
Regulated German label term indicating a quality wine made from late-harvest grapes of higher-than-average ripeness; the second lowest of the *Prädikat* levels, *Spätlese* wines are often lightly sweet and modest in alcohol, but some are heavier and drier.

Spritzy
Descriptive term for faintly carbonated wine.

Stainless steel
Material used in most modern fermentation vessels; in white wines, a descriptive term for an unoaked style.

Still
Term used for wine that has no carbonation or bubbles.

Strength
Descriptive term for wine's alcohol content. See **Weight**.

Subtle
Descriptive term for wines with lower-than-average intensity of olfactory scents and flavors.

Sweet, sweetness
Noticeable presence of sugar; one of the six true taste sensations detected by the tongue's taste buds.

Table grapes
Grapes grown for use as fresh produce, often seedless, juicy, and thin-skinned.

Tactile
Of or relating to the sense of touch.

Tannin
Phenolic compound with astringent properties found in grape skins; it acts as a natural preservative. The term "tannic" describes the dry feeling in the mouth after tasting red wines.

Taste, tasting
In common usage, all sensations derived from eating or drinking or the activity of sampling food or drink; for purposes of wine analysis, only the six sensations detectable with the tongue's taste buds.

Taste buds
The mechanism for perceiving taste sensations: clusters of nerves scattered across the tongue.

Tears
Better known as legs, these are the drips that form when wine is swirled in a glass.

Temperate
Climate category suitable for grape growing—neither too cold in winter nor too hot and tropical.

Terpene
Type of aromatic compound responsible for intense floral scents in grapes like Moscato, Gewurztraminer, and Riesling.

Terroir
Wine term for location-specific sensory characteristics in wine, often distinctive earthy aromas associated with a particular region or vineyard; may also refer to the unique aspects of a region or vineyard's geography that create these traits.

Texture
Descriptive term for wine's body or viscosity, typically driven by alcohol content. See **Weight**.

Toasted, toasty
Descriptive terms for wine with oak smells; nutty, caramelized scents and flavors derived from the flame "toasting" of wood during barrel-making.

Umami
One of the six sensations detected by the tongue's taste buds—an overall "yummy" taste caused by glutamates and amino acids.

Unoaked, unwooded, naked
Descriptive terms for wines that do not come into contact with oak or barrels during winemaking, or whose flavor and scent feature no detectable presence of new oak.

Unwooded
See **Unoaked**.

Vanillin
Principal flavor compound of vanilla beans, also strongly present in oak and a contributor of "oaky" flavors in wine.

Vin Doux Naturel
Type of French dessert wine made by *mutage*, or "Port method" fortification.

Vintage
The year in which a wine's grapes were harvested, often included on wine labels.

Vintner
Producer of wine.

Viscosity
Texture or thickness in a liquid; see **Weight**.

Vitis vinifera
Primary species of grapevine used for winemaking, of Eurasian origin.

Volatile
Evaporates readily at normal temperatures—a characteristic of many wine components, particularly alcohol and flavor compounds such as esters.

Weight
Descriptive term for wine's texture, perceived as thickness or viscosity in the mouth. Wines with more alcohol or lots of sugar feel heavier than those that are lower in alcohol and/or drier.

Winemaking
The process of transforming fresh grapes into wine through fermentation.

Workhorse grape
A grape variety capable of making pleasant wines even at very high yields; often used for bulk wines and bargain wines.

Yeast
Agent of fermentation essential for winemaking: single-celled microscopic organisms that convert sugar into alcohol.

Yield
Measure of vineyard productivity, typically in tons of grapes per acre or hectoliters of juice per hectare.

Young
Descriptive term for wine that is not barrel-aged before release or bottle-aged thereafter; typically applied to wines that are less than two years old.

INDEX

Page numbers in **bold** indicate
main entries

ABOUT THE AUTHOR

Sommelier **Marnie Old** is a breath of fresh air in the wine world, known for her engaging, intuitive, and refreshingly direct explanations of complex wine topics. Formerly the director of wine studies for Manhattan's esteemed French Culinary Institute, Marnie also served as the founding education chairperson for the American Sommelier Association. She pens a weekly wine column for the *Philadelphia Inquirer*, and her *Wine Simplified* series of video tutorials have millions of views on YouTube. Marnie's first book, also published by DK, was the popular *He Said Beer, She Said Wine*, an entertaining debate on food pairing coauthored with beer legend Sam Calagione, founder of Dogfish Head Craft Brewery. Marnie has also teamed up with wine legend Jean-Charles Boisset to coauthor their popular book *Passion for Wine*, and currently serves as Boisset Collection's director of *vin*lightenment.

AUTHOR'S ACKNOWLEDGMENTS

A number of people helped bring this book to life. I would like to extend my thanks and deepest appreciation to: Michael Mondavi, for his enduring support; Jamie Goode, for the vote of confidence; Tim Kilcullen, for his frank feedback; Karyn Gallagher, for her words of encouragement; Peggy Vance, for her intuitive grasp of the need for this book; Simon Murrell, for his brilliant visualizations; David Tombesi-Walton, for his patient attention to detail; Dawn Titmus for her editorial excellence; Glenda Fisher and Collette Sadler for their gorgeous graphics; David Ramey, for his last-minute fact-checking; Eric Miller, for his useful winemaking insights; Josh Rosenblat and Kaleigh Smith, for volunteering as guinea pigs; Kevin Zraly, for setting my feet on this path; Ewan Umholtz for three decades of inspiration and patience.

I would also like to thank the wine companies that have generously provided permission to reproduce images of their products:
Yellow Tail, Casella Wines, and Deutsch Family Wine & Spirits
Bodegas Muga and Jorge Ordóñez Selections
Seaglass Wines and Trinchero Family Estates
Ramey Wine Cellars
Argyle Winery
d'Arenberg and Old Bridge Cellars
Viña Concha y Toro
J. Moreau & Fils and Boisset Collection
Château La Lagune
Tenuta di Nozzole and Kobrand Corporation
Dr. Loosen and Loosen Bros USA
Pierre Gimonnet & Fils, Terry Theise, and Michael Skurnik Wines.

PUBLISHER'S ACKNOWLEDGMENTS

FIRST EDITION
Dorling Kindersley would like to thank the following people for their assistance with this project: Mandy Earey, Kathryn Wilding, and Kate Fenton for design; Elizabeth Clinton for editorial.

Sands Publishing Solutions would like to thank the following people for their input on this project: Natalie Godwin for design assistance; Hilary Bird for the index.

SECOND EDITION
Dorling Kindersley would like to thank Katie Hardwicke for proofreading.

PICTURE CREDITS

All images © Dorling Kindersley Limited, except p.187 Alamy Stock Photo: emer/Panther Media GmbH (bc); p.189 Alamy Stock Photo: Prisma by Dukas/Gian Carlo Patarino/Presseagentur GmbH (bl).

Penguin
Random
House

SECOND EDITION
Senior Editor Dawn Titmus
Senior Designer Glenda Fisher
US Editor Megan Douglass
Designer Collette Sadler
Senior Jacket Designer Nicola Powling
Jackets Coordinator Lucy Philpott
Production Editor David Almond
Production Controller Kariss Ainsworth
Managing Editor Ruth O'Rourke
Managing Art Editor Christine Keilty
Art Director Maxine Pedliham
Publishing Director Katie Cowan

DK INDIA
Assistant Editor Ankita Gupta
DTP Designer Manish Upreti
Senior DTP Designer Pushpak Tyagi
Managing Editor Soma B. Chowdhury
Pre-production Manager Sunil Sharma
Editorial Head Glenda Fernandes
Design Head Malavika Talukder

FIRST EDITION
Managing Editor Dawn Henderson
Managing Art Editor Christine Keilty
Senior Jackets Creative Nicola Powling
Production Editor Raymond Williams
Production Controller Oliver Jeffreys
Creative Technical Support Sonia Charbonnier
Publisher Peggy Vance
Design Director Peter Luff

Produced for Dorling Kindersley by **Sands Publishing Solutions**
Project Editor David Tombesi-Walton
Project Art Editor Simon Murrell

This American Edition, 2021
First American Edition, 2014
Published in the United States by DK Publishing
1450 Broadway, Suite 801, New York, NY 10018

Copyright © 2014, 2021 Dorling Kindersley Limited
DK, a Division of Penguin Random House LLC
21 22 23 24 25 10 9 8 7 6 5 4 3 2 1
001-323288–Oct/2021

A catalog record for this book
is available from the Library of Congress.
ISBN 978-0-7440-3986-3

DK books are available at special discounts when
purchased in bulk for sales promotions, premiums,
fund-raising, or educational use. For details, contact:
DK Publishing Special Markets,
1450 Broadway, Suite 801, New York, NY 10018
SpecialSales@dk.com

Printed and bound in UAE

For the curious

www.dk.com

This book was made with Forest Stewardship
Council ™ certified paper—one small step in
DK's commitment to a sustainable future.

For more information go to
www.dk.com/our-green-pledge